T0349930

THEORY OF
RELATIVITY

THEORY OF RELATIVITY

Fayyazuddin
National Centre for Physics, Pakistan

Riazuddin
National Centre for Physics, Pakistan

Muhammad Jamil Aslam
Quaid-i-Azam University, Pakistan

NEW JERSEY · LONDON · SINGAPORE · BEIJING · SHANGHAI · HONG KONG · TAIPEI · CHENNAI

Published by

World Scientific Publishing Co. Pte. Ltd.

5 Toh Tuck Link, Singapore 596224

USA office: 27 Warren Street, Suite 401-402, Hackensack, NJ 07601

UK office: 57 Shelton Street, Covent Garden, London WC2H 9HE

Library of Congress Cataloging-in-Publication Data

Fayyazuddin, 1930–   author.

   Theory of relativity / by Fayyazuddin (National Centre for Physics, Pakistan), Riazuddin (National Centre for Physics, Pakistan), Muhammad Jamil Aslam (Quaid-i-Azam University, Pakistan).

   pages cm

   Includes bibliographical references.

   ISBN 978-9814641890 (hardcover : alk. paper)

   1. Relativity (Physics) I. Riazuddin, author. II. Aslam, Jamil, author. III. Title.

QC173.55.F42 2015

530.11--dc23

                                                    2014050143

British Library Cataloguing-in-Publication Data

A catalogue record for this book is available from the British Library.

Printed in Singapore

To Our Children

Preface

By the turn of 20th century, the basic structure of physics, in which theories are formulated in terms of differential equations, predicting the future behavior in terms of states at a given instant of time was well-established. Lord Kelvin pointed out in the basic structure there are "two clouds at the horizon"; one is the failure to detect the existence of ether by the Michelson-Morlay experiment and the other is unable to use existing theory to account for the energy distribution of black body radiation. These two clouds led to the conceptual revolutions, Theory of Relativity and Quantum Mechanics.

The credit for the first revolution goes to Einstein. It changed our concept of space and time. Einstein formulated the special theory of relativity in 1905 and the general theory of relativity in 1911. In this theory, Einstein unified gravity with geometry. The formulation of Theory of relativity is the classical tradition unlike quantum mechanics.

The monograph covers both special and general theory of relativity. It is based on a course of lectures which the first two authors gave on the Theory of Relativity at the Punjab University Lahore, Quaid-i-Azam University Islamabad, The King Fahd University Dehran (R), The King Saud University Riyadh and The King Abdul Aziz University Jeddah (F).

The monograph is divided into three parts. The Special Theory of Relativity (Chapters 1 - 7); Space Time Groups including introduction to Supersymmetry (Chapter 8) and the General Theory of Relativity (Chapters 9 - 12). The first three chapters can be used by undergraduate students in their course on Classical Mechanics and Electrodynamics. Monograph itself can supplement graduate level courses on Electrodynamics, High Energy Physics, Quantum Field Theory, General Theory of Relativity and Cosmology.

We have tried to keep presentation simple and provide sufficient details in order to facilitate the understanding of the subject. The problems have been selected to clarify the presentation and solutions of selected problems are given for better understanding of the contents.

The first draft of monograph was complete before one of the authors, Riazuddin passed away. The remaining tasks were to include some new problems, add solutions to the problems and put the monograph in final form.

Finally we wish to thank Drs. Ishtiaq Ahmed and M. Ali Paracha who helped in typing the manuscript.

Fayyazuddin
M. Jamil Aslam
November 10, 2014

Contents

General Theory of Relativity: Riemannian Geometry; Curved Space Time 131

PART 1

Special Theory of Relativity

Chapter 1

GALILEAN TRANSFORMATIONS

1.1 Introduction

Both gravity and electromagnetism are long range forces and can be detected on a macroscopic scale and that is why they were the first to have
theories developed for them. Newton developed a theory of gravity in the
17th century and Maxwell did the same for electromagnetism in the 19th
century: The two theories are basically in conflict. Maxwell theory defined
a preferred velocity, the speed of light; whereas the Newtonian theory was
invariant if the whole system was given any uniform velocity. It turned
out that the Newtonian theory had to be modified to make it compatible
with Maxwell theory. This was achieved by Einstein in 1905 in his special theory of relativity and in 1917 for his general theory of relativity. In
the words of Einstein: "The special theory of relativity is an adaptation of
physical principles to Maxwell-Lorentz electrodynamics.... The postulate
of equivalence of inertial frames for the formulation of the laws of Nature is
assumed to be valid for the whole of physics (special relativity principle).
From Maxwell-Lorentz electrodynamics it takes the postulate of invariance
of the velocity of light in a vacuum (light principle)". Thus no information
can be transmitted faster than the speed of light. To harmonize the relativity principle with the light principle, the assumption that an absolute time
(agreeing for all inertial frames) exists, had to be abandoned. The laws of
transformation for space coordinates and time for the transition from one
intertial frame to another, the Lorentz transformations as they are termed,
are unequivocally established by these definitions. These transformations
leave invariant, the distance (ds) between two space-time points (t, x, y, z)
and $(t + dt, x + dx, y + dy, z + dz)$ defined by the equation

$$ds^2 = c^2 dt^2 - dx^2 - dy^2 - dz^2,$$

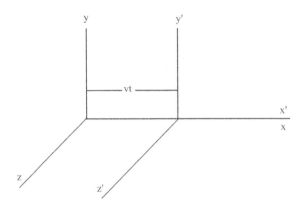

Fig. 1.1   The frames $S(x, y, z, t)$ and $S'(x', y', z', t')$ where S' is moving with respect to S with velocity v in the x-direction.

which can be measured by means of scales and clocks; x, y, z and t represent space and time coordinates and time with reference to a local inertial frame. Space and time coordinates are treated on an equal footing, resulting in the unification of space and time. According to the principle of relativistic invariance: All laws of physics take the same form in all inertial frames.

1.2   Galilean Transformations

Consider two frames $S(x, y, z, t)$ and $S'(x', y', z', t')$ where S' is moving with respect to S with velocity v in the x-direction as shown in Fig. 1.1. The Galilean transformation relates S and S' as follows

$$x' = x - vt$$
$$y' = y$$
$$z' = z$$
$$t' = t \tag{1.1}$$
$$\frac{dx'}{dt} = \frac{dx}{dt} - v$$
$$u' = u - v. \tag{1.2}$$

The velocities u' and u of a particle of mass m in the two frames are related by Eq. (1.2). Consider a simple collision between two particles moving in the x-direction. Momentum conservation in frame S gives

$$m_A u_A + m_B u_B = m_C u_C + m_D u_D.$$

Fig. 1.2   The collision of two particles moving in x-direction.

In moving frame S', the momentum conservation holds as well, i.e.

$$m_A u_A' + m_B u_B' = m_C u_C' + m_D u_D',$$

provided the mass is conserved

$$m_A + m_B = m_C + m_D.$$

This follows from Eq. (1.2)

$$m_A (u_A - v) + m_B (u_B - v) = m_C (u_C - v) + m_D (u_D - v),$$

$$m_A u_A + m_B u_B - (m_A + m_B) v = m_C u_C + m_D u_D - (m_C + m_D) v.$$

The law of inertia holds both in S and S' frames. Such frames are called the inertial frames.

1.3   Newtonian Mechanics and the Principle of Galilean Relativity

Newton's second law of motion is

$$m \frac{d^2 \mathbf{x}}{dt^2} = \mathbf{F}. \tag{1.3}$$

All reference frames in which Eq. (1.3) is valid are called inertial frames. If Eq. (1.3) is valid in reference frame S, it is also valid in a frame S', where S' and S are related to each other by the transformation

$$\mathbf{x}' = R\mathbf{x} - \mathbf{v}t,$$

$$t' = t, \tag{1.4}$$

where $\mathbf{v}$ is any real constant vector and R is any real constant orthogonal matrix. S' sees the S coordinate axes rotated by R, moving with velocity $-\mathbf{v}$. The transformations in Eq. (1.4) are called Galilean transformations. From Eq. (1.4)

$$\frac{d\mathbf{x}'}{dt'} = \frac{d\mathbf{x}'}{dt} \frac{dt}{dt'} = \frac{d\mathbf{x}'}{dt}$$

$$= R \frac{d\mathbf{x}}{dt} - \mathbf{v},$$

$$\frac{d^2 \mathbf{x}'}{dt'^2} = \frac{d^2 \mathbf{x}'}{dt^2} = R \frac{d^2 \mathbf{x}}{dt^2}.$$

Equation (1.3) can be written as

$$mR^{-1}\frac{d^2\mathbf{x}'}{dt'^2} = \mathbf{F},$$

$$m\frac{d^2\mathbf{x}'}{dt^2} = R\mathbf{F} = \mathbf{F}'. \tag{1.5}$$

In Newtonian mechanics m is an absolute quantity viz $m' = m$. $\mathbf{F}'$ gives the same force in the rotated coordinate system. From Eq. (1.5) one has the important result that Newton's laws of motion are invariant under Galilean transformations [Eq. (1.4)]. This is called the principle of Galilean relativity.

Example: Take $\mathbf{F}$ as the gravitational force between particles of mass m and M viz

$$\mathbf{F} = G\frac{mM\left(\mathbf{x}_M - \mathbf{x}_m\right)}{\left|\mathbf{x}_M - \mathbf{x}_m\right|^3}$$

$$= -GmM\frac{\mathbf{r}}{r^3}$$

$$= -\frac{GmM}{r^2}\mathbf{e}_r, \ \mathbf{e}_r = \frac{\mathbf{r}}{r}$$

where $\mathbf{r} = -\left(\mathbf{x}_M - \mathbf{x}_m\right)$ and $\mathbf{x}_M$ and $\mathbf{x}_m$ are the position vectors of M and m, respectively

$$R\mathbf{F} = -G\frac{mM}{r^2}R\mathbf{e}_r = -G\frac{mM}{r^2}\mathbf{e}_r' = \mathbf{F}'.$$

The equations of motion can hold in their usual form in only a limited class of coordinate systems called inertial frames [given in Eq. (1.4) with $\mathbf{v}$, R independent of t]. What then determines which reference frames are inertial frames? Newton answered that there must exist an absolute space, and that the inertial frames were those at rest in absolute space, or in a state of uniform motion with respect to absolute space.

1.3.1   *Electrodynamics and Galilean Relativity*

The theory of electrodynamics presented in 1864 by Maxwell does not satisfy the principle of Galilean relativity. For one thing, Maxwell's equations predict that the speed of light in vacuum is a universal constant c, but if this is true in a coordinate system $\mathbf{x}$, t, then it will not be true in the moving coordinate system $\mathbf{x}'$, t' defined by the Galilean transformation in (1.4). To see this, consider a plane electromagnetic wave travelling with a propagation vector $\mathbf{k}$ and frequency ω

$$\Psi\left(\mathbf{x},t\right) = A\sin\left(\omega t - \mathbf{k}\cdot\mathbf{x}\right). \tag{1.6}$$

Phase velocity is given by

$$\omega - \mathbf{k} \cdot \frac{d\mathbf{x}}{dt} = 0,$$

$$\mathbf{k} \cdot \mathbf{u} = \omega. \tag{1.7}$$

Phase velocity is along the direction of propagation $\mathbf{k}$. Let $\mathbf{n}$ be a unit vector in the direction of propagation:

$$\mathbf{k} = k\mathbf{n}, \mathbf{u} = c\mathbf{n}, ck = \omega,$$

$$c = \frac{\omega}{k} = \nu\lambda,$$

$$k = |\mathbf{k}|. \tag{1.8}$$

The inertial frames S and S' are related by Galilean transformation

$$t' = t, \tag{1.9}$$

$$\mathbf{x}' = \mathbf{x} - \mathbf{v}t. \tag{1.10}$$

The phase of a wave at the same physical point in two frames S and S' must be the same

$$\omega t - \mathbf{k} \cdot \mathbf{x} = \omega't' - \mathbf{k}' \cdot \mathbf{x}'. \tag{1.11}$$

Making use of the Galilean transformation, viz Eqs. (1.9) and (1.10), the above equation can be written as

$$\omega t - k\mathbf{n} \cdot \mathbf{x} = \omega't - k'\mathbf{n}' \cdot (\mathbf{x} - \mathbf{v}t).$$

Equating the coefficients of t and x give

$$\omega = \omega' + k'\mathbf{n}' \cdot \mathbf{v}, \tag{1.12}$$

$$k\mathbf{n} \cdot \mathbf{x} = k'\mathbf{n}' \cdot \mathbf{x}. \tag{1.13}$$

From Eq. (1.13), one obtains

$$k\mathbf{n} = k'\mathbf{n}', \tag{1.14}$$

$$k = k' \Rightarrow \frac{\omega}{c} = \frac{\omega'}{c'}, \tag{1.15}$$

since $n^2 = 1 = n'^2$. Using Eqs. (1.14) and (1.15), we get from Eq. (1.12)

$$\nu' = \nu\left(1 - \frac{\mathbf{n} \cdot \mathbf{v}}{c}\right). \tag{1.16}$$

Then from Eq. (1.15)

$$c' = c - \mathbf{n} \cdot \mathbf{v}, \tag{1.17}$$

$$\lambda' = \lambda, \tag{1.18}$$

which are in contradiction with experimental facts viz $c' = c$ and $\lambda' \neq \lambda$.

For the special case

$$\mathbf{v} = v\,(1,0,0), \tag{1.19}$$

$$\mathbf{n} = (\cos\theta, \sin\theta, 0), \tag{1.20}$$

$$\nu' = \nu\left(1 - \frac{v}{c}\cos\theta\right), \tag{1.21}$$

$$c' = c - v\cos\theta, \tag{1.22}$$

$c' \neq c$ is in conflict with Maxwell-Lorentz electrodynamics in which the speed of light in vacuum is constant. Hence Maxwell's equations do not satisfy the principle of Galilean relativity.

The question naturally arises:

Are Maxwell's equations valid only in the coordinate frames at rest with respect to the ether or is the principle of Galilean relativity to be modified.

All attempts to measure the velocity of earth with respect to ether failed, even though the earth has a velocity of 30 km/sec relative to the centre of our galaxy (Michelson-Morely experiment).

However Lorentz proved that instead of the Galilean transformation (1.1), Maxwell's equations are invariant under the transformations:

$$x' = \kappa\frac{\mathbf{x} - vt}{\sqrt{1 - \frac{v^2}{c^2}}}, y' = \kappa y, z' = \kappa z, t' = \frac{\left(t - \frac{v}{c^2}\right)x}{\sqrt{1 - \frac{v^2}{c^2}}}. \tag{1.23}$$

Poincaré extended the work of Lorentz regarding invariance of Maxwell's equations under "Lorentz transformations" (1.23). He also stated the principle of relativity "All laws of nature must be covariant with respect to Lorentz transformations." The work of Lorentz and Poincaré was on the basis of Maxwell's equations. Einstein formulated the new theory by stressing that covariance of laws of nature should follow from some basic postulates without reference to any dynamical law such as Maxwell's equations.

Chapter 2

LORENTZ TRANSFORMATIONS

2.1 The Two Postulates of Relativity

(1) The laws of physics take the same form in all inertial frames (Relativity principle).

(2) In any given inertial frame, the velocity of light c is the same whether the light be emitted by a body at rest or a body in uniform motion (light principle).

Consider two inertial frames S (x, y, z, t) and $S'(x', y', z', t')$. The S' frame is moving with velocity v in the x-direction with respect to S. At $t = t' = 0$, the two frames coincide and at that moment, a spherical light wave is emitted from the point at origin. t seconds later the wave has propagated to the surface of the sphere:

$$x^2 + y^2 + z^2 = c^2 t^2. \tag{2.1}$$

The compatibility of the two postulates demands that the corresponding equation in S' is

$$x'^2 + y'^2 + z'^2 = c^2 t'^2.$$

Assuming the relationship between two sets of coordinates to be linear in accordance with homogeneity of space and time we have

$$x'^2 + y'^2 + z'^2 - c^2 t'^2 = \kappa \left(x^2 + y^2 + z^2 - c^2 t^2 \right), \tag{2.2}$$

where κ is a constant depending on v. Noting that any motion parallel to x-axis must remain so after the transformation, we get from Eq. (2.2)

$$x' = \kappa \frac{x - vt}{\sqrt{1 - v^2/c^2}}, y' = \kappa y, z' = \kappa z, t' = \kappa \frac{\left(t - vx/c^2\right)}{\sqrt{1 - v^2/c^2}}. \tag{2.3}$$

We still have to show that

$$\kappa(v) = 1.$$

To do this, let us apply the transformation Eq. (2.3) once more, with the velocity in the opposite direction

$$x'' = \kappa(-v)\frac{x' + vt'}{\sqrt{1 - v^2/c^2}}, \quad y'' = \kappa(-v)y', \quad z'' = \kappa(-v)z', \quad t'' = \kappa(-v)\frac{(t' + vx'/c^2)}{\sqrt{1 - v^2/c^2}}. \tag{2.4}$$

Then, substituting Eq. (2.3) into Eq. (2.4), we have

$$x'' = \kappa(-v)\kappa(v)\frac{x - vt + v\left(t - vx/c^2\right)}{(1 - v^2/c^2)}$$

$$= \kappa(-v)\kappa(v)x,$$

$$y'' = \kappa(-v)\kappa(v)y, z'' = \kappa(-v)\kappa(v)z,$$

$$t'' = \kappa(-v)\kappa(v)\frac{t - vx/c^2 + (v/c^2)(x - vt)}{1 - v^2/c^2},$$

$$= \kappa(-v)\kappa(v)t. \tag{2.5}$$

Since S'' is at rest relative to S, it must be identical with it, hence

$$\kappa(v)\kappa(-v) = 1, \tag{2.6}$$

$\kappa(v)$ must be independent of the direction of v. The symmetry demands that the transformations on y and z should not change if $v \to -v$. Hence

$$\kappa(v) = \kappa(-v). \tag{2.7}$$

Therefore,

$$\kappa(v) = 1. \tag{2.8}$$

It is clear that we have used the group property of Lorentz transformations to derive Eq. (2.8). Hence we have the Lorentz transformation:

$$x' = \frac{x - vt}{\sqrt{1 - v^2/c^2}}; y = y'; z = z', t' = \frac{t - vx/c^2}{\sqrt{1 - v^2/c^2}}, \tag{2.9}$$

and

$$c^2t^2 - x^2 - y^2 - z^2 = c^2t'^2 - x'^2 - y'^2 - z'^2. \tag{2.10}$$

Thus in the theory of relativity $c^2t^2 - x^2 - y^2 - z^2$ is invariant resulting in the unification of space and time. For an arbitrary velocity $\mathbf{v}$, the transformations (2.9) take the form [$\mathbf{x} = x^1, x^2, x^3$]

$$\mathbf{x}' = \mathbf{x} + [(\gamma - 1)\frac{\mathbf{v}.\mathbf{x}}{v^2}\mathbf{v} - \gamma\mathbf{v}t], \quad t' = \gamma[t - \frac{\mathbf{v}.\mathbf{x}}{c^2}]. \tag{2.11}$$

Application 1: As a first application of Lorentz transformations, we re-consider the topic discussed in sec. 1.3.1. The phase of an electromagnetic wave should be the same at the same physical point in the two frames S and S'. i.e.,

$$\omega t - k \left(x \cos \theta + y \sin \theta\right) = \omega' t' - k' \left(x' \cos \theta' + y \sin \theta'\right), \tag{2.12}$$
$$= \gamma \omega' \left(t - \frac{v}{c^2} x\right) - k' \gamma \left(x - vt\right) \cos \theta' - k' y \sin \theta'.$$

From these equations, one gets

$$\omega = \gamma \left[\omega' + vk' \cos \theta'\right],$$
$$-k \cos \theta = \gamma \left[-\frac{v}{c^2} \omega' - k' \cos \theta'\right], \tag{2.13}$$
$$k \sin \theta = k' \sin \theta'.$$

Since velocity of light c is the same in the two frames viz

$$c = \frac{\omega}{k} = \frac{\omega'}{k'}, \tag{2.14}$$

the above equations become

$$\cos \theta' = \frac{\cos \theta - v/c}{1 - v/c \cos \theta}, \tag{2.15}$$

$$\sin \theta' = \frac{\sin \theta}{\gamma \left(1 - v/c \cos \theta\right)}, \tag{2.16}$$

$$\nu' = \gamma \nu \left(1 - \frac{v}{c} \cos \theta\right), \tag{2.17}$$

$$\lambda' = \frac{\lambda}{\gamma} \frac{1}{1 - v/c \cos \theta}. \tag{2.18}$$

Equations (2.17) and (2.18) give the correct expressions for the relativistic Doppler shift.

Application 2: As a second application of Lorentz transformation, we derive the law of addition of velocities. To do this, consider two frames S' and S'' moving in x-direction relative to S. Let the velocity of S' relative to S be v and the velocity of S'' relative to S' be u'. Let the velocity of S'' relative to S be $u'' = u$.

The Lorentz transformations give

$$x' = \gamma (v) (x - vt),$$
$$t' = \gamma (v) \left(t - \frac{vx}{c^2}\right), \tag{2.19}$$
$$x'' = \gamma (u') (x' - u't'),$$
$$t'' = \gamma (u') \left(t' - \frac{u'x'}{c^2}\right), \tag{2.20}$$

$$x'' = \gamma(u)(x - ut),$$
$$t'' = \gamma(u)\left(t - \frac{ux}{c^2}\right). \tag{2.21}$$

Substituting the values of x' and t' from Eq. (2.19) into Eq. (2.20), we get back Eq. (2.21) provided

$$\gamma(u) = \gamma(v)\gamma(u')\left(1 + \frac{vu'}{c^2}\right), \tag{2.22}$$

$$\gamma(u)u = \gamma(v)\gamma(u')(u' + v). \tag{2.23}$$

Hence, we have

$$u = \frac{\gamma(v)\gamma(u')(u' + v)}{\gamma(u)} = \frac{u' + v}{1 + vu'/c^2}, \tag{2.24}$$

to be compared with the Galilean transformation which gives $u = u' + v$. Equation (2.24) gives the law of addition of velocities in the theory of relativity. As an application of Eqs. (2.22) and (2.24), we consider the propagation of light in a moving medium (the drag coefficient). The velocity of light in a medium such as glass or water is given by c/n, where n is the refractive index of the medium.

Suppose the medium is moving with velocity v parallel to direction of light (frame S' is located in the medium). Relative to a stationary observer (i.e., an observer in frame S), the velocity of light is given by

$$u = \frac{u' + v}{1 + vu'/c^2} = \frac{c/n + v}{1 + (v/c^2)(c/n)}$$
$$= \frac{c}{n}\left(1 + \frac{vn}{c}\right)\left(1 + \frac{v}{nc}\right)^{-1},$$

$$u \approx \frac{c}{n} + v\left(1 - \frac{1}{n^2}\right) \text{ for } \frac{v^2}{c^2} \ll 1. \tag{2.25}$$

Thus the relativistic combination of velocities leads to drag coefficient without any extra assumption, to precisely the result that Fresnel and other theorists had to explain in terms of a partial dragging of the light by the medium.

2.2   Lorentz–FitzGerald Contraction

Consider a measuring rod which is at rest relative to S' and is placed parallel to x'-axis. S' is moving relative to S with velocity v along x-axis. Let l_0 be the length of the rod in S', i.e.,

$$l_0 = x'_2 - x'_1. \tag{2.26}$$

From the Lorentz transformation, we have

$$l_0 = x_2' - x_1' = \gamma(x_2 - vt) - \gamma(x_1 - vt)$$
$$= \gamma(x_2 - x_1) = \gamma l,$$
$$l = \frac{l_0}{\gamma} = \sqrt{1 - \frac{v^2}{c^2}} l_0. \tag{2.27}$$

In this derivation, we have defined the length l of the rod relative to S as the difference between simultaneous coordinate values of the end-points; by simultaneity we understand simultaneity relative to S. It is clear from Eq. (2.27) that the rod is contracted for the observer at rest. This is known as a phenomenon of length contraction.

2.3 Time Dilation

The most important and the simplest consequence of Lorentz transformation is the time dilation of moving clocks. To observer O'(system S') moving with the clock, the clock is at rest, but to observer O (system S) the clock is moving with velocity v along x-axis.

An observer looking at a clock at rest (system S' in which clock is at rest), will see two ticks separated by a space-time intervals

$$\Delta x' = 0, \quad \Delta t' = t_2' - t_1', \tag{2.28}$$

where $\Delta t'$ is the minimal period between ticks intended by the manufacturer. A second observer (in system S) who sees the clock moving with velocity v will observe that two ticks are separated by a time interval

$$\Delta t = t_2 - t_1 = \frac{t_2' + vx'/c^2}{\sqrt{1 - v^2/c^2}} - \frac{t_1' + vx'/c^2}{\sqrt{1 - v^2/c^2}},$$
$$= \frac{t_2' - t_1'}{\sqrt{1 - v^2/c^2}} = \frac{\Delta t'}{\sqrt{1 - v^2/c^2}} = \gamma \Delta t'. \tag{2.29}$$

Hence

$$\Delta t' < \Delta t, \tag{2.30}$$

i.e., clock which is moving relative to S will be slow compared with the clock in S for the observer O. It is commonly known as time dilation.

Equation (2.29) is literally being verified every day by experiments that measure the mean lifetime of rapidly moving unstable particles (for example: the μ-lepton) from cosmic rays or accelerators. Equation (2.29) tells us that a moving particle will have a mean life τ larger than the mean life τ_0, which it has at rest, by a factor γ viz

$$\tau = \frac{\tau_0}{\sqrt{1 - v^2/c^2}}.$$

This is in perfect agreement with experiment. In other words a rapidly moving particle lives longer in the laboratory frame.

We may use a radiating atom as a clock, the number of light waves emitted per second bring a measure of the rate of the atomic clock. If $\nu_0 = 1/\Delta t_0$ is the proper frequency of the atom, i.e., frequency of the emitted light measured in the frame S_0 in which the atom is at rest, then the frequency $1/\Delta t$ of the emitted light from an atom moving with velocity $\mathbf{v}$, as measured by a stationary observer, is given by

$$\frac{1}{\Delta t} = \frac{1}{\gamma \Delta t_0} = \frac{\nu_0}{\gamma}$$

$$= \sqrt{1 - \frac{v^2}{c^2}} \nu_0. \qquad (2.31)$$

However, during this time the distance from the observer to the light source will have increased by an amount $v_r \Delta t$, where v_r is the component of $\mathbf{v}$ along the direction from observer to light source (source is moving away from the observer). This causes the Doppler shift of the frequency. Thus the observed frequency ν_{obs} is given by

$$\nu_{obs} = \frac{1}{1 + v_r/c} \frac{1}{\Delta t}$$

$$= \left(1 + \frac{v_r}{c}\right)^{-1} \sqrt{1 - \frac{v^2}{c^2}} \nu_0$$

$$= \left(1 + \frac{v}{c}\cos\theta\right)^{-1} \sqrt{1 - \frac{v^2}{c^2}} \nu_0, \qquad (2.32)$$

$$\lambda_{obs} = \left(1 + \frac{v_r}{c}\right) \frac{1}{\sqrt{1 - v^2/c^2}} \lambda_0$$

$$= \left(1 + \frac{v}{c}\cos\theta\right) \frac{1}{\sqrt{1 - v^2/c^2}} \lambda_0. \qquad (2.33)$$

We have three cases:

(1) $v_r > 0$ light source is moving away $\lambda_{obs} > \lambda_0$, light is necessarily red shifted.

(2) $v_r = 0$, the light source is moving transversely

$$\lambda_{obs} = \frac{1}{\sqrt{1 - v^2/c^2}} \lambda_0, \lambda_{obs} > \lambda_0. \tag{2.34}$$

We have pure time dilation red shift, a purely relativistic effect.

(3) $v_r = -v$

$$\lambda_{obs} = \left(1 - \frac{v}{c}\right) \frac{1}{\sqrt{1 - v^2/c^2}} \lambda_0. \tag{2.35}$$

This gives violet shifted light. Let us apply Eq. (2.32) to far away galaxies which are receding from us. Suppose a light source located in a galaxy has a frequency ν_0. Equation (2.32) gives ($v_r = v$)

$$\nu_{obs} = \frac{\sqrt{1 - v^2/c^2}}{1 + v/c} \nu_0,$$

$$\lambda_{obs} = \sqrt{\frac{1 + v/c}{1 - v/c}} \lambda_0, \tag{2.36}$$

that is the red shifted $\lambda_{obs} > \lambda_0$.

2.4 Proper Time

The space-time interval

$$dx^2 + dy^2 + dz^2 - c^2 dt^2,$$

is Lorentz invariant, i.e.,

$$dx'^2 + dy'^2 + dz'^2 - c^2 dt'^2 = dx^2 + dy^2 + dz^2 - c^2 dt^2, \tag{2.37}$$

$$c\, d\tau'^2 = c^2 d\tau^2, \tag{2.38}$$

where

$$c\, d\tau = \left(c^2 dt^2 - d\mathbf{x}^2\right)^{\frac{1}{2}}, \tag{2.39}$$

τ is called the proper time and it is Lorentz invariant.

Now

$$\frac{d\tau^2}{dt^2} = 1 - \frac{1}{c^2} \left(\frac{d\mathbf{x}}{dt}\right)^2$$

$$= 1 - \frac{v^2}{c^2}. \tag{2.40}$$

Hence

$$dt = \gamma d\tau. \tag{2.41}$$

2.5   Transformation of Particle Velocities

Consider two inertial frames S and S'. For simplicity, consider S' to be moving relative to S with velocity v along x-axis. The motion of particle in S' frame is described by a trajectory

$$\mathbf{x}' = \mathbf{x}'(t'). \tag{2.42}$$

The velocity of the particle relative to S is given by

$$\mathbf{u} = \frac{d\mathbf{x}}{dt}. \tag{2.43}$$

The velocity of particle relative to S' is given by

$$\mathbf{u}' = \frac{d\mathbf{x}'}{dt'}. \tag{2.44}$$

From Lorentz transformations (2.11):

$$\mathbf{u}' = \frac{d\mathbf{x}'}{dt}\frac{dt}{dt'}$$
$$= \gamma^{-1}(1 - \frac{\mathbf{v}.\mathbf{u}}{c^2})^{-1}[\mathbf{u} + (\gamma - 1)\frac{\mathbf{v}.\mathbf{u}}{v^2}\mathbf{v} - \gamma\mathbf{v}].$$

For $\mathbf{v} = v\mathbf{e}_x$, where $\mathbf{e}_x$ is a unit vector along x-axis, one has

$$\mathbf{u}' = (\frac{1 - vu_x}{c^2})^{-1}\frac{1}{\gamma}[\mathbf{u} - (u_x - \gamma(u_x - v))\mathbf{e}_x)], \tag{2.45}$$

so that

$$u'_x = \left(\frac{1 - vu_x}{c^2}\right)^{-1}(u_x - v),$$

$$u'_y = \left(\frac{1 - vu_x}{c^2}\right)^{-1}\frac{1}{\gamma}u_y,$$

$$u'_z = \left(\frac{1 - vu_x}{c^2}\right)^{-1}\frac{1}{\gamma}u_z.$$

Chapter 3

RELATIVISTIC MECHANICS

3.1 Momentum and Energy

Consider a collision between two particles of masses m_1 and m_2 to form a third particle of mass M. This appears in frame S as shown in Fig. 3.1. In the frame S, momentum conservation gives

$$m_1 u = Mv, \tag{3.1}$$

where u and v are the velocities of particles of masses m_1 and M, respectively. The mass m_2 is taken to be at rest in frame S.

Fig. 3.1 m_1 is moving with velocity u and m_2 is taken to be at rest in frame S. After collision the mass M is moving with velocity v.

In the frame S', traveling with velocity v relative to S, the particle M is at rest and collision appears as shown in Fig. 3.2. Therefore, in the frame S'

$$m_1 u_1' + m_2 u_2' = 0. \tag{3.2}$$

In classical mechanics, the relations between the velocities are

$$u_1' = u - v, \tag{3.3}$$

$$u_2' = 0 - v = -v. \tag{3.4}$$

Therefore, Eq. (3.2) gives

$$m_1 (u - v) + m_2 (-v) = 0,$$

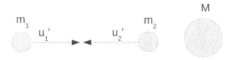

Fig. 3.2   Collision in frame S' which is moving with velocity v relative to S.

or

$$m_1 u = (m_1 + m_2)\, v. \tag{3.5}$$

Momentum conservation in both frames hold provided that mass is conserved viz

$$m_1 + m_2 = M. \tag{3.6}$$

In the theory of relativity

$$u'_1 = \frac{u-v}{1 - uv/c^2},$$

$$u'_2 = -v. \tag{3.7}$$

Using the values of u'_1 and u'_2 in Eq. (3.2), we have

$$m_1 u - (m_1 + m_2)\, v + m_2 \frac{uv^2}{c^2} = 0, \tag{3.8}$$

i.e., if momentum conservation in S' holds then it does not hold in S. Since S and S' are the inertial frames, the momentum conservation should hold in both frames. To accomplish this we have to modify the definition of momentum in special theory of relativity. Let us define momentum in the special theory of relativity as

$$p = m\gamma(u)u,$$

$$\gamma(u) = \frac{1}{\sqrt{1 - \frac{u^2}{c^2}}}. \tag{3.9}$$

In terms of new definitions, the momentum conservation in S frame gives

$$m_1 \gamma(u)\, u = M \gamma(v)\, v, \tag{3.10}$$

whereas in S' frame we have

$$m_1 \gamma(u'_1)\, u'_1 + m_2 \gamma(u'_2)\, u'_2 = 0. \tag{3.11}$$

Then using Eq. (3.7) and

$$\gamma(u'_1) = \gamma(u)\, \gamma(v) \left(1 - \frac{uv}{c^2}\right), \tag{3.12}$$

$$\gamma(u'_2) = \gamma(-v) = \gamma(v), \tag{3.13}$$

we get from Eq. (3.11)

$$m_1\gamma\left(u\right)\gamma\left(v\right)\left(u-v\right)-m_2\gamma\left(v\right)v=0. \tag{3.14}$$

Rewriting Eq. (3.14) in the form

$$m_1\gamma\left(u\right)u-M\gamma(v)v+[M\gamma(v)-m_1\gamma(u)-m_2]v=0,$$

we see that momentum conservation in S, viz Eq. (3.10), holds if

$$-m_1\gamma(u)-m_2+M\gamma(v)=0,$$

or

$$m_1\gamma(u)c^2+m_2c^2=M\gamma(v)c^2. \tag{3.15}$$

Each term in Eq. (3.15) has dimension of energy. Hence if one defines the relativistic energy as

$$E_1=m_1\gamma\left(u\right)c^2=\frac{m_1c^2}{\sqrt{1-\frac{u^2}{c^2}}}, \tag{3.16}$$

$$E_2=m_2c^2, \tag{3.17}$$

$$E_f=M\gamma\left(v\right)c^2=\frac{Mc^2}{\sqrt{1-\frac{v^2}{c^2}}}, \tag{3.18}$$

then Eq. (3.15) gives the energy conservation

$$E_1+E_2=E_f. \tag{3.19}$$

Thus momentum conservation in both the frames implies energy conservation in contrast to mass conservation relation (3.5) in Galilean relativity. Hence with the definitions

$$\mathbf{p}=\frac{m\mathbf{u}}{\sqrt{1-\frac{u^2}{c^2}}}=m\gamma(u)\mathbf{u}, \tag{3.20}$$

$$E=\frac{mc^2}{\sqrt{1-\frac{u^2}{c^2}}}=m\gamma(u)c^2, \tag{3.21}$$

$$\mathbf{u}=\frac{\mathbf{p}}{m\gamma(u)}=c^2\frac{\mathbf{p}}{E}, \tag{3.22}$$

momentum conservation holds in both inertial frames. The above equations define the momentum and energy in special theory of relativity. Also in terms of relativistic energy and momentum, we have a relation

$$E^2-c^2\mathbf{p}^2=m^2\frac{1}{1-\frac{u^2}{c^2}}\left[c^4-c^2u^2\right]=m^2c^4,$$

$$\left(\frac{E}{c}\right)^2-\mathbf{p}^2=m^2c^2, \tag{3.23}$$

i.e.,

$$\left(\frac{E}{c}\right)^2 - \mathbf{p}^2,$$

is relativistically invariant. In particular for a particle at rest ($\mathbf{p} = 0$), hence $E = mc^2$, the famous rest mass energy relation. We also note that

$$t^2 - \frac{x^2 + y^2 + z^2}{c^2} = t^2 - \frac{\mathbf{x}^2}{c^2},$$

is invariant. Thus $\frac{E}{c}$ and $\mathbf{p}$ transform in the same way as t and $\frac{\mathbf{x}}{c}$. Hence under Lorentz transformations in the x direction

$$p'_x = \gamma\left(p_x - \frac{v}{c^2}E\right) = p_x + [p_x(\gamma - 1) - \gamma\frac{v}{c^2}E],$$

$$p'_y = p_y,$$

$$p'_z = p_z,$$

$$\frac{E'}{c} = \gamma\left(\frac{E}{c} - \frac{v}{c^2}cp_x\right) \implies E' = \gamma(E - vp_x). \tag{3.24}$$

The above equations can be generalized to the case when the primed frame is moving w.r.t. unprimed frame with velocity $\mathbf{v}$

$$\mathbf{p}' = \mathbf{p} + \left[\mathbf{v}\frac{\mathbf{v}\cdot\mathbf{p}}{\mathbf{v}^2}(\gamma - 1) - \gamma\frac{\mathbf{v}}{c^2}E\right],$$

$$E' = \gamma[E - \mathbf{v}\cdot\mathbf{p}]. \tag{3.25}$$

If we write $\mathbf{p} = (\mathbf{p}_\perp, \mathbf{p}_\parallel)$, where $\mathbf{p}_\parallel$ is the component of $\mathbf{p}$ along $\mathbf{v}$, then we have

$$\mathbf{p}'_\parallel = \gamma\left[\mathbf{p}_\parallel - \frac{\mathbf{v}}{c^2}E\right],$$

$$\mathbf{p}'_\perp = \mathbf{p}_\perp, \tag{3.26}$$

$$E' = \gamma[E - vp_\parallel].$$

We define the longitudinal rapidity η of particle as

$$\tanh\eta = \frac{v}{c} = \frac{cp_\parallel}{E}, \tag{3.27}$$

$$\eta = \tanh^{-1}\frac{cp_\parallel}{E},$$

$$= \frac{1}{2}\ln\frac{E + cp_\parallel}{E - cp_\parallel}. \tag{3.28}$$

Noting from Eq. (3.26),

$$E' + cp'_\| = \gamma\left(1 - \frac{v}{c}\right)(E + cp_\|),$$

$$E' - cp'_\| = \gamma\left(1 + \frac{v}{c}\right)(E - cp_\|). \tag{3.29}$$

Thus

$$E'^2 - c^2 p'^2_\| = E^2 - c^2 p^2_\| = m_T^2 c^4,$$

$$E = m_T c^2 \cosh\eta,$$

$$cp_\| = m_T c^2 \sinh\eta,$$

and

$$\eta' = \eta + \frac{1}{2}\ln\frac{1 - v/c}{1 + v/c},$$

$$d\eta' = d\eta.$$

Thus $d\eta$ is Lorentz invariant, i.e., the shape of rapidity distribution $\frac{dN}{d\eta}$ is invariant.

The transformation law for Newtonian force can be derived from Eq. (3.25)

$$\mathbf{F}' = \frac{d\mathbf{p}'}{dt'} = \frac{dt}{dt'}\frac{d\mathbf{p}'}{dt},$$

$$\frac{dt'}{dt} = \gamma[1 - \frac{\mathbf{v}\cdot\mathbf{u}}{c^2}],$$

where $\mathbf{u} = d\mathbf{x}/dt$ is the velocity of a particle. Thus

$$\mathbf{F}' = \frac{1}{\gamma}[1 - \frac{\mathbf{v}.\mathbf{u}}{c^2}]^{-1}\{\mathbf{F} + (\gamma - 1)\frac{\mathbf{F}.\mathbf{v}}{v^2}\mathbf{v} - \gamma\frac{\mathbf{v}}{c^2}\frac{dE}{dt}\}.$$

From Eqs. (3.23) and (3.22)

$$\frac{E}{c^2}\frac{dE}{dt} = \mathbf{p}\cdot\frac{d\mathbf{p}}{dt},$$

$$\frac{dE}{dt} = \mathbf{u}\cdot\frac{d\mathbf{p}}{dt} = \mathbf{u}\cdot\mathbf{F},$$

so that

$$\mathbf{F}' = \frac{1}{\gamma}[1 - \frac{\mathbf{v}.\mathbf{u}}{c^2}]^{-1}\{\mathbf{F} + (\gamma - 1)\frac{\mathbf{F}.\mathbf{v}}{v^2}\mathbf{v} - \gamma\frac{\mathbf{u}.\mathbf{F}}{c^2}\mathbf{v}\}. \tag{3.30}$$

3.2   Application of Relativistic Mechanics

3.2.1   *Doppler Shift*

Consider a stationary atom or a nucleus. If it is in an excited state A^*, it would emit a photon and goes over to stable atom A:

$$A^* \to A + \gamma.$$

Let M be the mass of A^* and M' be that of A. Then if we neglect the recoil, we have

$$Mc^2 = M'c^2 + E_\gamma^0 = M'c^2 + h\nu_0 \tag{3.31}$$

where ν_0 is the frequency of emitted photon. Now taking into account the recoil, energy-momentum conservation gives

$$Mc^2 = E_f + E_\gamma = E_f + h\nu, \tag{3.32}$$

$$0 = \mathbf{p}_f + \frac{E_\gamma}{c}\mathbf{n}, \tag{3.33}$$

$$\mathbf{p}_f^2 = \frac{E_\gamma^2}{c^2},$$

$$\left(E_f^2 - M'^2 c^4\right) = E_\gamma^2. \tag{3.34}$$

Using Eqs. (3.32) and (3.34), one gets:

$$\left(Mc^2 - E_\gamma\right)^2 = E_f^2 = M'^2 c^4 + E_\gamma^2,$$

then by using Eq. (3.31), we obtain

$$2Mc^2 E_\gamma = M^2 c^4 - M'^2 c^4 = M^2 c^4 - \left(Mc^2 - E_\gamma^0\right)^2.$$

Finally:

$$E_\gamma = E_\gamma^0 \left(1 - \frac{E_\gamma^0}{2Mc^2}\right),$$

$$h\nu = h\nu_0 \left(1 - \frac{h\nu_0}{2Mc^2}\right). \tag{3.35}$$

Consider now the emission of a photon, when the atom A^* is moving with velocity $\mathbf{v}$.

The energy momentum conservation gives

$$E = E' + E_\gamma', \tag{3.36}$$

$$\mathbf{p} = \mathbf{p}' + \frac{E_\gamma'}{c}\mathbf{n}, \tag{3.37}$$

$$\mathbf{p}'^2 = \mathbf{p}^2 + \frac{E_\gamma'^2}{c^2} - 2\frac{E_\gamma'}{c}\mathbf{p}\cdot\mathbf{n},$$

where **n** is a unit vector. Thus

$$E'^2 - M'^2 c^4 = E^2 - M^2 c^4 + E_\gamma'^2 - 2cE_\gamma' p \cos\theta,$$

with θ an angle between **p** and **n**. Using Eq. (3.36) and

$$v = \frac{c^2 p}{E}, \tag{3.38}$$

we get

$$M^2 c^4 - M'^2 c^4 = 2EE_\gamma' \left(1 - \frac{v}{c}\cos\theta\right). \tag{3.39}$$

Then using Eq. (3.32) and

$$E = M\gamma c^2, \tag{3.40}$$

we obtain

$$2Mc^2 E_\gamma^0 - E_\gamma^{02} = 2M\gamma c^2 E_\gamma' \left(1 - \frac{v}{c}\cos\theta\right). \tag{3.41}$$

Making use of Eq. (3.35) gives

$$E_\gamma' = \frac{1}{\gamma} \frac{E_\gamma}{1 - \frac{v}{c}\cos\theta}, \tag{3.42}$$

or in terms of frequency

$$\nu' = \frac{\nu \left(1 - v^2/c^2\right)^{\frac{1}{2}}}{1 - v/c \cos\theta}. \tag{3.43}$$

We have obtained the Doppler shift Eqs. (2.32) and (2.33) from relativistic mechanics. Note that ν' is the observed frequency, whereas ν is the proper frequency of the atom.

3.3 Scattering Kinematics

3.3.1 *Two-Particle Scattering*

$$a + b \rightarrow c + d.$$

For the scattering process, energy momentum is conserved:

$$p_a + p_b = p_c + p_d,$$

$$\mathbf{p}_a + \mathbf{p}_b = \mathbf{p}_c + \mathbf{p}_d,$$

$$E_a + E_b = E_c + E_d. \tag{3.44}$$

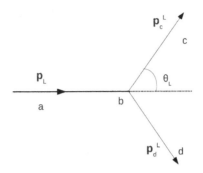

Fig. 3.3   Scattering in the Lab. frame.

We can form the following Lorentz scalars

$$s = (p_a + p_b)^2 = (p_c + p_d)^2,$$

$$t = (p_a - p_c)^2 = (p_d - p_b)^2,$$

$$u = (p_a - p_d)^2 = (p_c - p_b)^2, \tag{3.45}$$

where s, t and u are called Mandelstam variables. By adding these quantities we get

$$s + t + u = m_a^2 + m_b^2 + m_c^2 + m_d^2. \tag{3.46}$$

We will take (s,t) as independent variables. The invariants s, u, t are frame independent. Two frames are relevant: **Lab Frame (Fig. 3.3)**

$$p_a = \left(E_a^L,\ \mathbf{p}_a^L\right) = \left(\nu_L, \mathbf{p}_L\right),$$

$$p_b = (m_b, 0),$$

$$p_c = \left(E_c^L,\ \mathbf{p}_c^L\right), p_d = \left(E_d^L,\ \mathbf{p}_d^L\right),$$

$$s = p_a^2 + p_b^2 + 2p_a \cdot p_b = m_a^2 + m_b^2 + 2m_b\nu_L, \tag{3.47}$$

$$t = p_a^2 + p_c^2 - 2p_a \cdot p_c = m_a^2 + m_c^2 - 2\nu_L E_c^L + 2\left|\mathbf{p}_a^L\right|\left|\mathbf{p}_c^L\right|\cos\theta_L, \tag{3.48}$$

$$\nu_L = \frac{s - m_a^2 - m_b^2}{2m_b}, \tag{3.49}$$

$$\mathbf{p}_L^2 = \nu_L^2 - m_a^2 = \frac{\left[s - \left(m_a^2 + m_b^2\right)\right]^2}{4m_b^2} - m_a^2,$$

$$|p_L| = \sqrt{\frac{\lambda\left(s, m_a^2, m_b^2\right)}{2m_b}}, \tag{3.50}$$

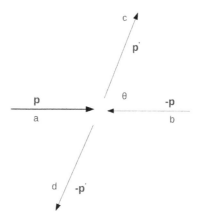

Fig. 3.4   Two-body scattering in the centre of mass frame.

where

$$\lambda\left(s, m_a^2, m_b^2\right) = s^2 + m_a^4 + m_b^4 - 2sm_a^2 - 2sm_b^2 - 2m_a^2 m_b^2,$$

$$\lambda\left(x, y, z\right) = x^2 + y^2 + z^2 - 2xy - 2xz - 2yz. \tag{3.51}$$

Centre of Mass Frame: (Fig. 3.4)

$$p_a = \left(E_a, \mathbf{p}\right),$$

$$p_b = \left(E_b, -\mathbf{p}\right),$$

$$p_c = \left(E_c, \mathbf{p}'\right),$$

$$p_d = \left(E_d, -\mathbf{p}'\right), \tag{3.52}$$

where

$$s = p_a^2 + p_b^2 + 2p_a \cdot p_b = m_a^2 + m_b^2 + 2E_a E_b + 2\mathbf{p}^2$$

$$= \left(E_a + E_b\right)^2$$

$$= \left(E_c + E_d\right)^2 = E_{cm}^2, \tag{3.53}$$

$$t = p_a^2 + p_c^2 - 2p_a \cdot p_c$$

$$= m_a^2 + m_c^2 - 2E_a E_c + 2\left|\mathbf{p}\right|\left|\mathbf{p}'\right|\cos\theta$$

$$= m_b^2 + m_d^2 - 2E_b E_d + 2\left|\mathbf{p}\right|\left|\mathbf{p}'\right|\cos\theta.$$

For elastic scattering

$$a + b = a + b,$$

$$\left|\mathbf{p}\right| = \left|\mathbf{p}'\right|,$$

$$t = -2\left|\mathbf{p}\right|^2\left(1 - \cos\theta\right). \tag{3.54}$$

In general

$$s = (E_a + E_b)^2 = \left(\sqrt{\mathbf{p}^2 + m_a^2} + \sqrt{\mathbf{p}^2 + m_b^2}\right)^2$$

$$= \mathbf{p}^2 + m_a^2 + \mathbf{p}^2 + m_b^2 + 2\sqrt{\mathbf{p}^2 + m_a^2}\sqrt{\mathbf{p}^2 + m_b^2},$$

which gives

$$4s\mathbf{p}^2 = s^2 + m_a^4 + m_b^4 - 2sm_a^2 - 2sm_b^2 - 2m_a^2 m_b^2,$$

$$|\mathbf{p}| = \sqrt{\frac{\lambda(s, m_a^2, m_b^2)}{2\sqrt{s}}}, \tag{3.55}$$

$$|\mathbf{p}'| = \sqrt{\frac{\lambda(s, m_c^2, m_d^2)}{2\sqrt{s}}}, \tag{3.56}$$

$$E_a = \sqrt{\mathbf{p}_a^2 + m_a^2} = \sqrt{\mathbf{p}^2 + m_a^2} = \frac{s + m_a^2 - m_b^2}{2\sqrt{s}}, \tag{3.57}$$

$$E_b = \frac{s + m_b^2 - m_a^2}{2\sqrt{s}}, \tag{3.58}$$

$$E_c = \frac{s + m_c^2 - m_d^2}{2\sqrt{s}}, \tag{3.59}$$

$$E_d = \frac{s + m_d^2 - m_c^2}{2\sqrt{s}}. \tag{3.60}$$

The centre of mass frame and the lab. frame are related to each other by Lorentz transformation. Centre of mass frame is moving with respect to lab. frame with velocity $\mathbf{v}$:

$$\mathbf{v} = \frac{\mathbf{p}_L}{\nu_L + m_b}. \tag{3.61}$$

Conversely lab. frame is moving with respect to c.m. frame with velocity $-\mathbf{v}$. For $\mathbf{p}_L$ along z-axis

$$p_L = \gamma \left[p + vE_a\right],$$

$$\nu_L = \gamma \left[E_a + vp\right],$$

$$m_b = \gamma \left[E_b - vp\right]. \tag{3.62}$$

These equations give

$$\gamma = \frac{\nu_L + m_b}{E_{cm}}. \tag{3.63}$$

Furthermore, the Lorentz transformation gives

$$p_c^L \cos \theta_L = \gamma \left[p' \cos \theta + v E_c \right],$$

$$p_c^L \sin \theta_L = p' \sin \theta,$$

$$E_c^L = \gamma \left[E_c + v p' \cos \theta \right]. \tag{3.64}$$

Hence

$$\tan \theta_L = \frac{p' \sin \theta}{\gamma \left[p' \cos \theta + v E_c \right]}. \tag{3.65}$$

3.4 Motion of a Charged Particle in a Uniform Magnetic Field

The equation of motion of a charged particle in electromagnetic field is given by the Lorentz equation

$$\frac{d\mathbf{p}}{dt} = e \left[\varepsilon + \frac{1}{c} \mathbf{v} \times \mathbf{B} \right], \tag{3.66}$$

where ε is the electric field, $\mathbf{v}$ is the velocity of a particle and $\mathbf{B}$ is the applied magnetic field. Now

$$\mathbf{p} = m\gamma \mathbf{v}, \quad E = m\gamma c^2,$$

$$\frac{d\mathbf{p}}{dt} = m\gamma \frac{d\mathbf{v}}{dt} + m\mathbf{v} \frac{d\gamma}{dt}, \tag{3.67}$$

$$\mathbf{v} \cdot \frac{d\mathbf{p}}{dt} = m \frac{\gamma}{2} \frac{d(\mathbf{v} \cdot \mathbf{v})}{dt} + mv^2 \frac{d\gamma}{dt} = mc^2 \frac{d\gamma}{dt},$$

where we have used

$$\frac{1}{2} \gamma \frac{dv^2}{dt} = c^2 \left(1 - \frac{v^2}{c^2} \right) \frac{d\gamma}{dt}.$$

On the other hand, from Eq. (3.66)

$$\mathbf{v} \cdot \frac{d\mathbf{p}}{dt} = e\mathbf{v} \cdot \varepsilon,$$

so that (cf. Eq. (3.67))

$$\frac{d\gamma}{dt} = \frac{e}{mc^2} \mathbf{v} \cdot \varepsilon. \tag{3.68}$$

Thus from Eq. (3.67), using (3.68), we get

$$\frac{d\mathbf{v}}{dt} = \frac{e}{m\gamma} [\varepsilon + \frac{1}{c} \mathbf{v} \times \mathbf{B} - \frac{\mathbf{v}}{c^2} \mathbf{v} \cdot \varepsilon].$$

We consider the case when $\varepsilon = 0$. Then

$$\mathbf{F} = \frac{d\mathbf{p}}{dt} = \frac{e}{c}\mathbf{v} \times \mathbf{B}. \qquad (3.69)$$

Now

$$\mathbf{v} \cdot \mathbf{F} = \frac{e}{c}\mathbf{v} \cdot (\mathbf{v} \times \mathbf{B}) = 0. \qquad (3.70)$$

Hence energy of the particle is not changed when it is moving in a uniform magnetic field, i.e.,

$$\frac{dE}{dt} = 0. \qquad (3.71)$$

This means that the magnitude of the velocity $\mathbf{v}$ is not changed but its direction changes continuously. Hence it describes a circle of constant radius. We take the magnetic field $\mathbf{B}$ perpendicular to $\mathbf{v}$. Let us take $\mathbf{B}$ along z-axis so that $\mathbf{v}$ is in the x-y plane. In the x-y plane, we use the polar coordinates ρ and ϕ. Thus

$$\mathbf{r} = z\hat{e}_z + \rho\left(\cos\phi\,\hat{e}_x + \sin\phi\,\hat{e}_y\right),$$

$$\mathbf{v} = \frac{d\mathbf{r}}{dt} = \frac{dz}{dt}\hat{e}_z + \rho\left(-\sin\phi\frac{d\phi}{dt}\hat{e}_x + \cos\phi\frac{d\phi}{dt}\hat{e}_y\right).$$

Now we take $\frac{dz}{dt} = 0$ so that

$$\mathbf{v} = \rho\omega\left[-\sin\phi\,\hat{e}_x + \cos\phi\,\hat{e}_y\right], \qquad (3.72)$$

where

$$\omega = \frac{d\phi}{dt} \qquad (3.73)$$

is the angular velocity. Then

$$\frac{d\mathbf{v}}{dt} = -\rho\omega^2\left[\cos\phi\,\hat{e}_x + \sin\phi\,\hat{e}_y\right]$$

$$= -\rho\omega^2\hat{\rho}, \qquad (3.74)$$

where $\hat{\rho}$ is a unit vector along the radius vector. Hence from Eq. (3.69), we get

$$m\gamma\frac{d\mathbf{v}}{dt} = \frac{e}{c}\mathbf{v} \times \mathbf{B}$$

$$= \frac{eB}{c}\rho\omega\left[\sin\phi\,\hat{e}_y + \cos\phi\,\hat{e}_x\right]$$

$$= -\frac{eB}{c}\rho\omega\hat{\rho}. \qquad (3.75)$$

Comparing Eqs. (3.74) and (3.75), we get the magnitude of the angular velocity

$$\omega = \frac{eB}{\gamma mc},$$

$$v = \rho\omega = \frac{eB\rho}{\gamma mc},$$

$$m\gamma v = \frac{eB\rho}{c},$$

$$p = \frac{eB}{c}\rho. \tag{3.76}$$

Thus the particle describes a circle of radius $\rho = cp/eB$. This is the result which one also gets from Newtonian mechanics. By measuring the radius of curvature of a charged particle moving in a known magnetic field, one can determine the momentum of a charged particle. The above result is of immense use in the detection techniques of particle physics and also in cyclotron.

Now we consider the motion in a constant electric field ε. For this case we take $\mathbf{B} = 0$. The equation of motion is given by

$$\frac{d\mathbf{p}}{dt} = e\varepsilon, \tag{3.77}$$

we consider the motion in the x-y plane and take ε along the y-axis. At $t = 0$, we take

$$p_x = p_0, \ p_y = 0. \tag{3.78}$$

Now we have

$$\frac{dp_x}{dt} = 0 \Rightarrow p_x = p_0,$$

$$\frac{dp_y}{dt} = e\varepsilon \Rightarrow p_y = e\varepsilon t. \tag{3.79}$$

The energy

$$E^2 = m^2c^4 + c^2 \left(p_x^2 + p_y^2\right) = m^2c^4 + c^2p_0^2 + c^2 \left(e\varepsilon t\right)^2$$

$$= E_0^2 + c^2e^2\varepsilon^2t^2, \tag{3.80}$$

where

$$E_0^2 = m^2c^4 + c^2p_0^2. \tag{3.81}$$

Now using the relation

$$E = \frac{mc^2}{\sqrt{1 - \frac{v^2}{c^2}}}. \tag{3.82}$$

We get from Eq. (3.79)

$$\frac{dx}{dt} = \frac{p_0}{m}\sqrt{1 - \frac{v^2}{c^2}} = \frac{c^2 p_0}{\sqrt{E_0^2 + c^2 e^2 \varepsilon^2 t^2}}, \tag{3.83}$$

$$\frac{dy}{dt} = \frac{c^2 e \varepsilon t}{\sqrt{E_0^2 + c^2 e^2 \varepsilon^2 t^2}}. \tag{3.84}$$

Integrating Eqs. (3.83) and (3.84), with the boundary conditions at $t = 0$, $x = 0 = y$, we get

$$x = \frac{c p_0}{e \varepsilon} \sinh^{-1}\left(\frac{c e \varepsilon t}{E_0}\right), \tag{3.85}$$

$$y = \frac{E_0}{e \varepsilon} - \frac{\sqrt{E_0^2 + c^2 e^2 \varepsilon^2 t^2}}{e \varepsilon}. \tag{3.86}$$

Eliminating t, between Eqs. (3.85) and (3.86), we get the trajectory of the charged particle moving in a uniform electric field

$$y = \frac{E_0}{e \varepsilon}\left[1 - \cosh\left(\frac{e \varepsilon x}{c p_0}\right)\right]. \tag{3.87}$$

In the non-relativistic limit, expanding $\cosh\left(\frac{e \varepsilon x}{c p_0}\right)$, in powers of $1/c$, and $E_0 = mc^2$

$$y = \frac{mc^2}{2e\varepsilon} \frac{e^2 \varepsilon^2 x^2}{c^2 p_0^2} = \frac{e \varepsilon m}{2 p_0^2} x^2, \tag{3.88}$$

i.e., the trajectory is a parabola, a result well known in classical mechanics.

3.5 Problems Related to Chapters 2 and 3

3.1 Consider two frames S_1 and S_2 moving in x-direction relative to frame S with velocities v_1 and v_2 respectively. Time interval measured in S in the clock located in S_1 is t. Show that corresponding time interval t_2 measured in S_2 is given by

$$t_2 = \gamma(v_2)\left(1 - \frac{v_1 v_2}{c^2}\right)t,$$

Solution

Let v_2' be the velocity of S_2 relative to S_1. We have (Eq. (2.24))

$$v_2 = \frac{v_2' + v_1}{1 + \frac{v_1 v_2'}{c^2}}.$$

From the above equation

$$v_2' = \frac{v_2 - v_1}{1 - \frac{v_1 v_2}{c^2}},$$

t_1: proper time of S_1

$$t = \gamma(v_1)t_1,$$

$$t_2 = \gamma(-v_2')t_1 = \frac{\gamma(v_2')}{\gamma(v_1)}t.$$

Now

$$1 - \frac{v_2'^2}{c^2} = 1 - \frac{(v_2 - v_1)^2}{(1 - \frac{v_1 v_2}{c^2})^2},$$

$$= \frac{(1 - \frac{v_1^2}{c^2})(1 - \frac{v_2^2}{c^2})}{(1 - \frac{v_1 v_2}{c^2})^2}.$$

Thus

$$\gamma(v_2') = \gamma(v_1)\gamma(v_2)(1 - \frac{v_1 v_2}{c^2}).$$

Hence

$$t_2 = \gamma(v_2)(1 - \frac{v_1 v_2}{c^2})t.$$

3.2 A particle is travelling with velocity u in the $x - y$ plane along a trajectory which makes an angle θ with the x-axis in frame S. Show that in frame S' which is moving relative to S with uniform velocity v along x-axis, the angle θ' is given by

$$\tan \theta' = \frac{1}{\gamma}(\frac{u \sin \theta}{u \cos \theta - v}).$$

Show that for a rod which is stationary in the frame S making an angle θ with x-axis, the angle in S' is given by

$$\tan \theta' = \gamma \tan \theta.$$

3.3 A photon of energy ε collides with another photon of energy E. Show that the minimum value of ε permitting the formation of pair of particles of mass m is

$$\varepsilon_{th} = \frac{2m^2 c^4}{E(1 - \cos \theta)},$$

where θ is an angle between two photons.

3.4 A K meson of mass m_K comes to rest and decays into a μ-meson of mass m_μ and a neutrino ν_μ whose mass can be taken to be zero

$$K^+ \to \mu^+ + \nu_\mu.$$

Find the energy of ν_μ and the velocity of μ-meson in terms of masses m_K and m_μ. μ-meson has a mean life time $\tau_0 \simeq 2.2 \times 10^{-6}$ sec. How much distance does μ-meson travel before it decays? Is the following reaction allowed?

$$\nu_\mu + p \to \mu^+ + n$$

where p and n denote proton and neutron, respectively. Take

$$m_K = 493.68 \frac{\text{MeV}}{c^2}, \ m_\mu = 105.66 \frac{\text{MeV}}{c^2}, \ m_p = 938.27 \frac{\text{MeV}}{c^2}, \ m_n = 939.56 \frac{\text{MeV}}{c^2}.$$

Solution:

$$K^+ \to \mu^+ + \nu_\mu.$$

Energy-momentum conservation

$$m_K c^2 = E_\mu + E_\nu, \tag{3.89}$$

$$0 = \mathbf{p}_\mu + \mathbf{p}_\nu,$$

$$E_\nu = c\,|\mathbf{p}_\nu|,$$

$$\mathbf{p}_\mu^2 = \mathbf{p}_\nu^2 = \frac{E_\nu^2}{c^2}, \tag{3.90}$$

from Eqs. (3.89) and (3.90)

$$m_K c^2 = E_\mu + E_\nu = E_\nu + \sqrt{E_\nu^2 + m_\mu^2 c^4}.$$

Thus we have

$$E_\nu = \frac{(m_K^2 - m_\mu^2)c^4}{2 m_K c^2} = 235.5 \text{ MeV},$$

$$E_\mu = \frac{(m_K^2 + m_\mu^2)c^4}{2 m_K c^2}.$$

Velocity of μ:

$$\frac{v}{c} = \frac{c\,|\mathbf{p}_\mu|}{E_\mu} = \frac{E_\nu}{E_\mu}$$

$$= \frac{m_K^2 - m_\mu^2}{m_K^2 + m_\mu^2}$$

$$\approx 0.914,$$

$$\sqrt{1 - \frac{v^2}{c^2}} = \frac{2m_K m_\mu}{m_K^2 + m_\mu^2},$$

$$\tau = \frac{\tau_0}{\sqrt{1 - \frac{v^2}{c^2}}}$$

$$= \frac{m_K^2 + m_\mu^2}{2m_K m_\mu}\tau_0$$

$$= \frac{1}{2}\frac{m_K}{m_\mu}(1 + \frac{m_\mu^2}{m_K^2})\tau_0$$

$$\approx 5.37 \times 10^{-6} \text{ sec.}$$

The distance travels before μ^+ decay:

$$v\tau = \frac{v}{c}\tau c \approx 1516 \text{ m.}$$

$$\nu_\mu + p \rightarrow \mu^+ + n$$

$$p_1 = (E_\nu/c, \mathbf{p}_\nu), \quad p_2 = (m_p c^2, 0),$$

p_1 and p_2 are four momenta of ν_μ and p, respectively.

$$s = (p_1 + p_2)^2 = m_p^2 c^2 + 2m_p E_\nu.$$

Threshold for this reaction

$$s_0 = (m_\mu^2 + m_n^2)c^2.$$

Thus for the reaction to occur

$$(m_p^2 c^2 + 2m_p E_\nu) \geq (m_\mu^2 + m_n^2)c^2$$

$$E_\nu \geq \frac{[m_\mu^2 + (m_n^2 - m_p^2)]c^2}{m_p^2 c^2} \approx 7.24 \text{ MeV.}$$

Hence the reaction is allowed.

3.5 (i) π^0 meson travelling with velocity v along x-axis decays into two γ-rays. Obviously in its rest frame, the angular distribution $P(\theta_c)$ is isotropic. Show that in the laboratory frame, angular distribution is given by

$$P(\theta) = \frac{(1 - \frac{v^2}{c^2})}{4\pi(1 - \frac{v}{c}\cos\theta)^2},$$

where θ is the angle as measured in the laboratory frame with respect to x-axis.

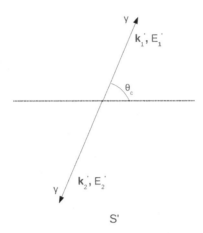

Fig. 3.5   Centre of mass frame.

(ii) Show that the frequency of the photon emitted at an angle θ with x-axis is given by

$$\nu_1 = \frac{1}{\gamma}\nu_0\left[\frac{1}{(1 - \frac{v}{c}\cos\theta)}\right] \text{ (Doppler shift)}$$

where $h\nu_0 = \frac{1}{2}m_{\pi^0}c^2$

(iii) If the energy of the π^0 is 6 GeV, how far will it travel before it decays. The mean life time of π^0 is $\tau_0 \simeq 8.4 \times 10^{-17}$ sec, $m_{\pi^0} = 134.98\frac{\text{MeV}}{c^2}$.

Solution

(i) In the rest frame S' of π^0 (cm frame)

$$E_1' = E_2' = E', \ \mathbf{k}_1' = -\mathbf{k}_2' = \mathbf{k}'$$

$$m_\pi c^2 = 2E'.$$

Obviously angular distribution is isotropic. In the Lab. frame S:

$$|\mathbf{k}_1| = E_1/c, \ |\mathbf{k}_2| = E_2/c.$$

Frame S' is moving with velocity v along x-axis. Therefore

$$k_{1x}' = k_1'\cos\theta_c = \gamma\left(k_1\cos\theta - \frac{v}{c^2}E_1\right)$$

or

$$E'\cos\theta_c = \gamma\left(E_1\cos\theta - \frac{v}{c}E_1\right) \tag{3.91}$$

$$E_1' = E' = \gamma\left(E_1 - \frac{vE_1}{c}\cos\theta\right). \tag{3.92}$$

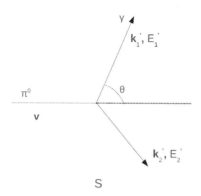

Fig. 3.6   Lab. frame S.

Hence

$$\cos\theta_c = \frac{\cos\theta - v/c}{1 - \frac{v}{c}\cos\theta}.$$

(3.93)

Now

$$P(\theta)d\Omega = P'(\theta_c)d\Omega_c = \frac{1}{4\pi}d\Omega_c.$$

From Eq. (3.93)

$$\sin\theta_c d\theta_c = \frac{\left(1 - v^2/c^2\right)\sin\theta d\theta}{\left(1 - \frac{v}{c}\cos\theta\right)^2}.$$

Thus

$$d\Omega_c = \frac{\left(1 - v^2/c^2\right)d\Omega}{\left(1 - \frac{v}{c}\cos\theta\right)^2}.$$

Hence

$$P\left(\theta\right) = \frac{1}{4\pi}\frac{\left(1 - v^2/c^2\right)}{\left(1 - \frac{v}{c}\cos\theta\right)^2}.$$

(ii) From Eq. (3.92)

$$E_1 = \frac{1}{\gamma}\frac{E'}{1 - \frac{v}{c}\cos\theta}.$$

Therefore

$$\nu_1 = \frac{1}{\gamma}\frac{\nu_0}{1 - \frac{v}{c}\cos\theta} \qquad \text{(Doppler Shift)}$$

where

$$h\nu_0 = E' = \frac{1}{2}m_{\pi^0}c^2.$$

(iii) Now $v/c = cp_\pi/E_\pi$. Thus

$$\sqrt{1 - \frac{v^2}{c^2}} = \frac{m_{\pi^0}c^2}{E_{\pi^0}}, \quad \frac{v}{c} = \left[1 - \left(\frac{m_{\pi^0}c^2}{E_{\pi^0}}\right)^2\right]^{1/2} \approx 1$$

$$\gamma = \frac{1}{\sqrt{1 - \frac{v^2}{c^2}}} = \frac{E_{\pi^0}}{m_{\pi^0}c^2} \approx 44.$$

Distance travelled by π^0 before it decays is

$$\gamma\tau_0 c \approx 1.11 \times 10^{-4} \text{ cm}.$$

3.6 For a nonrelativistic harmonic oscillator we have well-known relation $\tau = \frac{2\pi}{\omega}$. Show that for a relativistic simple harmonic oscillator, the above relation is modified to

$$\tau = \frac{2\pi}{\omega}[1 + \frac{3a^2\omega^2}{16c^2} + O(\frac{\omega^2}{c^2})^2]$$

where a is the amplitude of oscillation.

3.7 A neutron travelling with constant velocity v along the x-axis decay

$$n \rightarrow p + e^- + \bar{\nu}_e$$

with a mean life time $\tau_o = 880$ s. In the rest frame of neutron (c.m. frame S'), the electron comes with velocity u' making an angle θ' with x-axis:

a): Show that in the laboratory frame:
(i):

$$\tan\theta = \frac{1}{\gamma}\frac{u'\sin\theta'}{u'\cos\theta + v}, \quad \gamma = \gamma(v) = \frac{1}{\sqrt{1 - v^2/c^2}}.$$

(ii):

$$u_x = \frac{u'_x + v}{1 + \frac{vu'_x}{c^2}}$$

$$u_y = \frac{u'_y}{\gamma\left(1 + \frac{vu'_x}{c^2}\right)}.$$

Conversely

$$\tan \theta' = \frac{1}{\gamma} \frac{u \sin \theta}{u \cos \theta - v}$$

$$u'_x = \frac{u_x - v}{1 - \frac{vu_x}{c^2}}$$

$$u'_y = \frac{u_y}{\gamma \left(1 - \frac{vu_x}{c^2}\right)}.$$

b): Neutrons are produced in primary cosmic rays interactions in outer space. Suppose they are produced one light year from the earth. If half of neutrons are detected at the earth, show that the energy of the neutron produced in the cosmic rays interaction is 5×10^4 GeV.

Solution: a) (i): Let **k** be the momentum of electron in the Lab. frame and **k**$'$ is its momentum in the rest frame of the neutron. Neutron is travelling in $x - y$ plane. Thus

$$k_x = k \cos \theta = \gamma \left(k' \cos \theta' + \frac{v}{c^2} E' \right) \qquad (3.94)$$

$$k_y = k \sin \theta = k' \sin \theta' \qquad (3.95)$$

$$E = \gamma \left(E' + vk' \cos \theta' \right). \qquad (3.96)$$

Now

$$k = m\gamma(u) u, \quad k' = m\gamma(u') u'$$

$$E = m\gamma(u) c^2, \quad E' = m\gamma(u') c^2.$$

Therefore, from Eqs. (3.94) and (3.95), we have

$$\tan \theta = \frac{k' \sin \theta'}{\gamma \left(k' \cos \theta' + \frac{v}{c^2} E' \right)} = \frac{u' \sin \theta'}{\gamma \left(u' \cos \theta' + v \right)}.$$

From Eq. (3.96)

$$\gamma(u) = \gamma\gamma(u') \left(1 + \frac{vu'}{c^2} \cos \theta \right) = \gamma\gamma(u') \left(1 + \frac{vu'_x}{c^2} \right).$$

Hence from Eqs. (3.94) and (3.95)

$$u_x = \frac{\gamma\gamma(u')}{\gamma(u)} \left(u'_x + v \right) = \frac{u'_x + v}{\left(1 + \frac{vu'_x}{c^2} \right)}$$

$$u_y = \frac{\gamma(u')}{\gamma(u)} u'_y = \frac{u'_y}{\gamma \left(1 + \frac{vu'_x}{c^2} \right)}.$$

Similarly, one can derive the converse relations. Note that S' is travelling with velocity relative to S, where S is traveling relative to S' with velocity $-v$.

b): One light year = distance traveled by light in one year, i.e.,

$$ct = c\,(365 \times 24 \times 3600) = 0.945 \times 10^{16} \text{ m}$$

$$t_l = \frac{0.945 \times 10^6 m}{c} = 0.315 \times 10^8 \text{ sec.}$$

Now

$$N\,(t) = N_o e^{-t_l/\tau}$$

$$\frac{1}{2} = e^{-t_l/\tau} \rightarrow \tau = \frac{t_l}{\ln 2} = 0.454 \times 10^8 \text{ sec}$$

$$\frac{v_n}{c} = \frac{c p_n}{E_n}$$

$$\tau = \frac{\tau_o}{\sqrt{1 - \frac{v_n^2}{c^2}}} = \tau_o \frac{E_n}{m_n c^2}.$$

Hence

$$E_n = \tau \frac{m_n c^2}{\tau_o} \approx 5 \times 10^4 \text{ GeV.}$$

Chapter 4

LORENTZ TRANSFORMATIONS
(General Case)

4.1 Lorentz Transformations (General Case)(in Four Dimensional Space Time)

In the special theory of relativity, light principle states that the speed of light as measured by all observers is the same, i.e.,

$$c^2 t^2 - x^2 - y^2 - z^2 = c^2 t'^2 - x'^2 - y'^2 - z'^2,$$

$$c^2 t^2 - \mathbf{x}^2 = c^2 t'^2 - \mathbf{x}'^2 \; (invariant). \tag{4.1}$$

It is useful to introduce a 4-vector x^μ

$$x^\mu = \left(x^0, x^i \right) = (ct, \mathbf{x}), \tag{4.2}$$

called contravariant vector. We define another 4-vector x_μ

$$x_\mu = (x_0, x_i) = (ct, -\mathbf{x}), \tag{4.3}$$

called covariant vector. Equation (4.1) can be written as

$$x'^\mu x'_\mu = x^\mu x_\mu, \tag{4.4}$$

i.e., $x^\mu x_\mu$ is invariant.

A Lorentz transformation is a transformation from one system of space-time coordinates x^μ to another system x'^μ which leaves $x^\mu x_\mu$ invariant

$$x'^\mu = \sum_\nu \Lambda^\mu{}_\nu x^\nu = \Lambda^\mu{}_\nu \, x^\nu. \tag{4.5}$$

It is useful to introduce metric tensor

$$\eta^{\mu\nu} = \begin{pmatrix} 1 & 0 & 0 & 0 \\ 0 & -1 & 0 & 0 \\ 0 & 0 & -1 & 0 \\ 0 & 0 & 0 & -1 \end{pmatrix} \tag{4.6}$$

with

$$\eta_{\mu\nu}\eta^{\nu\lambda} = \delta^{\lambda}_{\mu}, \quad \det\eta = -1,$$

and

$$\eta_{00} = \eta^{00} = 1, \eta_{ij} = \eta^{ij} = -\delta_{ij}. \tag{4.7}$$

We note that the contravariant and covariant vectors are related to each through metric tensor, i.e.,

$$x^{\mu} = \eta^{\mu\nu}x_{\nu},$$

$$x_{\mu} = \eta_{\mu\nu}x^{\nu}, \tag{4.8}$$

$$x^{\mu}x_{\mu} = \eta^{\mu\lambda}x_{\lambda}x_{\mu}$$

$$= \eta_{\mu\lambda}x^{\mu}x^{\lambda}. \tag{4.9}$$

Invariance of $x^{\mu}x_{\mu}$ gives

$$x^{\mu}\eta_{\mu\nu}x^{\nu} = x'^{\alpha}\eta_{\alpha\beta}x'^{\beta}$$

$$= \Lambda^{\alpha}{}_{\mu}x^{\mu}\Lambda^{\beta}{}_{\nu}x^{\nu}\eta_{\alpha\beta}.$$

This implies

$$\eta_{\mu\nu} = \Lambda^{\alpha}{}_{\mu}\eta_{\alpha\beta}\Lambda^{\beta}{}_{\nu} = \left(\Lambda^{T}\right)_{\mu}{}^{\alpha}\eta_{\alpha\beta}\Lambda^{\beta}{}_{\nu} \tag{4.10}$$

$$= \left(\Lambda^{T}\eta\Lambda\right)_{\mu\nu}$$

or in matrix form

$$\Lambda^{T}\eta\Lambda = \eta, \tag{4.11}$$

$$\det(\Lambda^{T}\eta\Lambda) = \det\eta,$$

$$\det\eta(\det\Lambda^{2}) = \det\eta,$$

$$\det\Lambda = \pm 1. \tag{4.12}$$

If $\det\Lambda = 1$, Lorentz transformation (LT) is called proper, denoted by L_{+}. This excludes spatial reflection for which $\det\Lambda = -1$. We restrict to those transformations which are continuously connected to the identity transformation for which $\det\Lambda = 1$, i.e., to proper Lorentz transformations. Further setting $\mu = 0 = \nu$ in Eq. (4.10)

$$1 = \Lambda^{\alpha}{}_{0}\eta_{\alpha\beta}\Lambda^{\beta}{}_{0},$$

$$= (\Lambda^{0}{}_{0})^{2} - \sum_{i}(\Lambda^{i}{}_{0})^{2}.$$

Thus

$$(\Lambda^0{}_0)^2 = 1 + \sum_i (\Lambda^i{}_0)^2,$$

$$\geq 1.$$

Hence

$$\Lambda^0{}_0 \geq 1 \text{ or } \Lambda^0{}_0 \leq -1. \tag{4.13}$$

If $\Lambda^0{}_0 \geq 1$, the time direction is unaltered and it is called "Orthochronous", denoted by $L^\uparrow$. Since $\Lambda^0{}_0 = 1$ for the identity transformation, continuity requires that all proper Lorentz transformation have

$$\Lambda^0{}_0 \geq 1. \tag{4.14}$$

We shall restrict ourselves to $L^\uparrow_+$, which excludes space reflection and time reversal. For Lorentz transformations Λ_1 and Λ_2 obeying the constraint

$$(\Lambda_1 \Lambda_2)^T \eta (\Lambda_1 \Lambda_2) = \Lambda_2^T \Lambda_1^T \eta \Lambda_1 \Lambda_2$$

$$= \Lambda_2^T \eta \Lambda_2 = \eta, \tag{4.15}$$

the group property is satisfied.

We define

$$\partial^\mu = \frac{\partial}{\partial x_\mu} = (\frac{1}{c}\frac{\partial}{\partial t}, -\nabla),$$

$$\partial_\mu = \frac{\partial}{\partial x^\mu} = (\frac{1}{c}\frac{\partial}{\partial t}, \nabla).$$

For an infinitesimal Lorentz transformation

$$x'^\mu = (\delta^\mu_\nu + \epsilon^\mu{}_\nu) x^\nu = x^\mu + \epsilon^\mu{}_\nu x^\nu, \tag{4.16}$$

where

$$\Lambda^\mu{}_\nu = \delta^\mu_\nu + \epsilon^\mu{}_\nu \tag{4.17}$$

and the constraint (4.10) gives

$$\eta_{\mu\nu} = (\delta^\lambda_\mu + \epsilon^\lambda{}_\mu) \eta_{\lambda\sigma} (\delta^\sigma_\nu + \epsilon^\sigma{}_\nu) \tag{4.18}$$

$$= \eta_{\mu\nu} + \eta_{\mu\sigma} \epsilon^\sigma{}_\nu + \eta_{\nu\lambda} \epsilon^\lambda{}_\mu,$$

$$\epsilon_{\mu\nu} + \epsilon_{\nu\mu} = 0,$$

$$\epsilon_{\mu\nu} = -\epsilon_{\nu\mu}, \tag{4.19}$$

i.e., $\epsilon_{\mu\nu}$ is an antisymmetric tensor.

Now

$$\epsilon^\mu_{\ \nu} = \eta^{\mu\lambda}\epsilon_{\lambda\nu},$$

$$\epsilon_{\mu\nu} = \eta_{\mu\lambda}\,\epsilon^\lambda_{\ \nu}, \tag{4.20}$$

so that Eq. (4.16):

$$x'^\mu = x^\mu + \eta^{\mu\lambda}\epsilon_{\lambda\nu}x^\nu.$$

Thus for infinitesimal Lorentz transformation

$$x'^0 = x^0 + \epsilon_{0j}x^j,$$

$$= x^0 - \eta^j x^j = x^0 - \eta.\mathbf{x}, \tag{4.21}$$

$$x'^i = x^i + \eta^{ij}(\epsilon_{j0}x^0 + \epsilon_{jk}x^k),$$

$$= x^i - \epsilon_{i0}x^0 - \epsilon_{ik}x^k,$$

$$= x^i - \eta^i x^0 + \epsilon^{ikl}x^k\omega^l,$$

$$\mathbf{x}' = \mathbf{x} - \eta x^0 - (\omega \times \mathbf{x}), \tag{4.22}$$

where we have used

$$\epsilon_{0j} = \eta_j = -\eta^j$$

and

$$\epsilon_{ik} = \varepsilon_{ikl}x^l = -\varepsilon^{ikl}x^l. \tag{4.23}$$

Let us consider the special case viz

$$\omega = \omega\mathbf{e}_z,$$

$$\eta = \eta\mathbf{e}_x, \tag{4.24}$$

then from Eqs. (4.21) and (4.22), one gets

$$x'^1 = x^1 - \eta x^0 - \omega x^2,$$

$$x'^2 = x^2 - \omega x^1,$$

$$x'^3 = x^3,$$

$$x'^0 = x^0 - \eta x^1. \tag{4.25}$$

For a finite transformation, replace

$$1 \to \cos\omega, \omega \to \sin\omega,$$

$$1 \to \cosh\eta = \gamma; \eta \to \sinh\eta = \gamma\frac{v}{c}. \tag{4.26}$$

In terms of the above finite transformation, Eq. (4.25) takes the form

$$x'^1 = x^1 \cos\omega \cosh\eta - x^0 \cos\omega \sinh\eta + x^2 \sin\omega$$

$$= \gamma \left(x^1 - \frac{v}{c} x^0 \right) \cos\omega + x^2 \sin\omega$$

$$= x^1 \cos\omega + x^2 \sin\omega + v \cos\omega \left[(\gamma - 1) \frac{vx^1}{v^2} - \frac{\gamma}{c} x^0 \right],$$

$$x'^2 = -x^1 \sin\omega + x^2 \cos\omega,$$

$$x'^3 = x^3,$$

$$x'^0 = x^0 \cosh\eta - x^1 \sinh\eta = \gamma \left[x^0 - \frac{vx^1}{c} \right]. \tag{4.27}$$

It follows from Eqs. (4.27) that for $v = 0$, we have a rotation in 3-space, implying that rotation group is a subgroup of Lorentz group. The transformation that changes the velocity of the coordinate frame is Lorentz boost. If we put $\omega = 0$, we have pure Lorentz transformation in agreement with Eq. (2.9). The generalization of Eqs. (4.27)

$$x'^i = \delta^{ij} x^j + (\gamma - 1) \frac{v^i x^j}{v^2} v^j - \gamma \frac{v^i}{c} x^0, \tag{4.28}$$

$$x'^0 = \gamma x^0 - \gamma \frac{v^j x^j}{c}. \tag{4.29}$$

Remember: Unlike the rotation group SO(3), elements of Lorentz group can be unbounded, as the range of η is $(-\infty, \infty)$, Therefore, the Lorentz group is non compact.

4.2 Light Cone

With respect to an arbitrarily chosen coordinate origin, spacetime is divided into three distinct regions separated by the light cone which is defined by the equation

$$x^2 = c^2 t^2 - \mathbf{x}^2 = 0.$$

Future cone consists of all points with $x^2 > 0$, $x^0 > 0$, which for past cone $x^2 > 0, x^0 < 0$. The region outside the light cone is characterized by $x^2 < 0$. If we write the Lorentz transformation as

$$x'^\mu = \Lambda^\mu{}_\nu x^\nu$$

$$x'^0 = \Lambda^0{}_0 x^0 + \Lambda^0{}_j x^j$$

$$x'^i = \Lambda^i{}_0 x^0 + \Lambda^i{}_j x^j. \tag{4.30}$$

Then comparison with Eqs. (4.28) and (4.29) gives

$$\Lambda^0{}_0 = \gamma, \Lambda^0{}_i = -\gamma \frac{v^i}{c}, \Lambda^i{}_j = \delta^{ij} + \frac{v^i v^j}{v^2}(\gamma - 1), \ \Lambda^i{}_0 = -\gamma \frac{v^i}{c}. \quad (4.31)$$

We have already seen in Chapter 3 that $\frac{E}{c}$ and $\mathbf{p}$ transform in the same way as ct and $\mathbf{x}$ and $(\frac{E}{c})^2 - \mathbf{p}^2 = m^2 c^2$ is invariant just as $c^2 t^2 - \mathbf{x}^2$ is. Thus p^μ forms a 4-vector like x^μ.

$$p^\mu = (\frac{E}{c}, \mathbf{p}),$$

$$p'^\mu = \Lambda^\mu{}_\nu p^\nu.$$

4.3   Lorentz Boost Transformation

Let us find a Lorentz velocity transformations which "boosts" a particle from rest with 4-momentum $p_0^\mu = (mc, \mathbf{0})$ to the momentum $\mathbf{p}$:

$$p^\mu = \Lambda^\mu{}_\nu(\mathbf{v})p_0^\nu = L_\nu^\mu(p)p_0^\nu, \quad (4.32)$$

where $\Lambda^\mu{}_\nu(\mathbf{v})$ is give in Eq. (4.31). By using

$$v^i = c^2 \frac{p^i}{E}, \ \ E^2 - c^2 |\mathbf{p}|^2 = m^2 c^4,$$

we have

$$L_0^0 = \gamma = \frac{1}{\sqrt{1 - \frac{c^2 |\mathbf{p}|^2}{E^2}}} = \frac{E}{mc^2},$$

$$L_j^0 = \gamma \frac{v^j}{c} = \frac{p^j}{mc} = L_0^j,$$

$$L_j^i = \delta^{ij} + \frac{p^i p^j c^2}{E^2 - m^2 c^4}(\frac{E}{mc^2} - 1) = \delta^{ij} + \frac{p^i p^j c^2}{(E + mc^2) mc^2}. \quad (4.33)$$

We can write Lorentz boost operator in manifest covariant form [noting that $p_0^0(p + p_0)^0 = mc(\frac{E}{c} + mc)$, $p_0^i = 0$]

$$L_\lambda^\mu(p) = \eta_{\lambda\nu} L^{\mu\nu}(p),$$

$$L^{\mu\nu}(p) = \eta^{\mu\nu} - \frac{(p + p_0)^\mu (p + p_0)^\nu}{p_0 \cdot (p + p_0)} + \frac{2p^\mu p^\nu}{m^2 c^2}. \quad (4.34)$$

Note that $L^{\mu\nu}(p) \longrightarrow \eta^{\mu\nu}$ as $p_0^\mu \longrightarrow p_0^0$.

4.4 Vectors and Tensors

A contravariant 4-vector A^μ transforms like x^μ

$$A'^\mu = \Lambda^\mu{}_\nu A^\nu, \tag{4.35}$$

whereas a covariant 4-vector A_μ transforms as x_μ

$$A'_\mu = \Lambda_\mu{}^\nu A_\nu. \tag{4.36}$$

The norm of a 4-vector is given by

$$A^2 = A^\mu A_\mu = A^0 A_0 + A^i A_i = \left(A^0\right)^2 - \mathbf{A} \cdot \mathbf{A}, \tag{4.37}$$

it is invariant

$$A'^2 = A'^\mu A'_\mu = A^\alpha \eta_{\alpha\beta} A'^\beta = \Lambda^\alpha{}_\mu A^\mu \Lambda^\beta{}_\nu A^\nu \eta_{\alpha\beta}$$
$$= A^\mu A_\mu = A^2, \tag{4.38}$$

since

$$\Lambda^\alpha{}_\mu \eta_{\alpha\beta} \Lambda^\beta{}_\nu = \eta_{\mu\nu}, \tag{4.39}$$

$$\Lambda^T \eta \Lambda = \eta. \tag{4.40}$$

The scalar product

$$A \cdot B = A^\mu B_\mu$$

is clearly invariant. Now

$$\partial'^\mu = \frac{\partial}{\partial x'_\mu} = \frac{\partial}{\partial x_\sigma} \frac{\partial x_\sigma}{\partial x'_\mu},$$

but

$$x' = \Lambda x,$$

therefore,

$$x = \Lambda^T x',$$

$$x_\mu = \left(\Lambda^T\right)_\mu{}^\sigma x'_\sigma = \Lambda^\sigma{}_\mu x'_\sigma,$$

$$\frac{\partial x_\mu}{\partial x'_\rho} = \Lambda^\sigma{}_\mu \delta_{\sigma\rho} = \Lambda^\rho{}_\mu.$$

Hence

$$\partial'^\mu = \Lambda^\mu{}_\sigma \frac{\partial}{\partial x_\sigma} = \Lambda^\mu{}_\sigma \partial^\sigma, \tag{4.41}$$

transforms as a contravariant vector. Similarly, one can show that

$$\partial'_\mu = \Lambda_\mu{}^\nu \partial_\nu, \tag{4.42}$$

transforms as a covariant vector. We note that

$$\partial'^{\mu}\partial'_{\mu} = \partial^{\mu}\partial_{\mu} = \frac{1}{c^2}\frac{\partial^2}{\partial t^2} - \nabla^2, \tag{4.43}$$

is invariant.

A second rank contravariant tensor $T^{\mu\nu}$ transforms as

$$T'^{\mu\nu} = \Lambda^{\mu}{}_{\sigma}\,\Lambda^{\nu}{}_{\lambda}\,T^{\sigma\lambda}, \tag{4.44}$$

whereas a covariant second rank tensor $T_{\mu\nu}$ transforms as

$$T'_{\mu\nu} = \Lambda_{\mu}{}^{\sigma}\,\Lambda_{\nu}{}^{\lambda}\,T_{\sigma\lambda}. \tag{4.45}$$

A second rank mixed tensor transforms as

$$T'^{\mu}{}_{\nu} = \Lambda^{\mu}{}_{\sigma}\,\Lambda_{\nu}{}^{\lambda}\,T^{\sigma}_{\lambda}. \tag{4.46}$$

Now

$$\begin{aligned}
\partial^{\mu}A_{\mu} \rightarrow \partial'^{\mu}A'_{\mu} &= \Lambda^{\mu}{}_{\sigma}\,\partial^{\sigma}\,\Lambda_{\mu}{}^{\lambda}\,\eta_{\lambda\alpha}A^{\alpha} \\
&= \Lambda^{\lambda}{}_{\mu}\eta_{\lambda\alpha}\Lambda^{\mu}{}_{\sigma}\partial^{\sigma}A^{\alpha} \\
&= \eta_{\alpha\sigma}\partial^{\sigma}A^{\alpha} = \partial^{\sigma}A_{\sigma} = \partial^{\mu}A_{\mu},
\end{aligned} \tag{4.47}$$

is a scalar, called 4-divergence.

The second rank tensor $T^{\mu\nu}$ has 16 components. It can be expressed as

$$\begin{aligned}
T^{\mu\nu} &= \frac{1}{2}\left(T^{\mu\nu} + T^{\mu\nu}\right) + \frac{1}{2}\left(T^{\mu\nu} - T^{\mu\nu}\right) \\
&= S_{\mu\nu} + A_{\mu\nu},
\end{aligned}$$

where $S_{\mu\nu} = S_{\nu\mu}$ having 10 components is a symmetric tensor and $A_{\mu\nu} = -A_{\nu\mu}$ having 6 components is antisymmetric tensor.

The metric tensor

$$\eta^{\mu\nu} = \eta^{\nu\mu},$$

is called the Minkowski tensor

$$\eta'^{\mu\nu} = \Lambda^{\mu}{}_{\sigma}\,\Lambda^{\nu}{}_{\lambda}\,\eta^{\sigma\lambda} = \eta^{\mu\nu},$$

since $\eta'^{\mu\nu}$ is numerically the same as $\eta^{\mu\nu}$. The Levi-Civita tensor, totally antisymmetric in all indices, is defined as

$$\epsilon^{\mu\nu\rho\sigma} = \begin{cases} +1 & \text{if } \mu\nu\rho\sigma \text{ is an even permutation} \\ -1 & \text{if } \mu\nu\rho\sigma \text{ is an odd permutation} \\ 0 & \text{otherwise.} \end{cases} \tag{4.48}$$

Now

$$\epsilon_{0123} = \eta_{0\alpha}\eta_{1\beta}\eta_{2\gamma}\eta_{3\delta}\epsilon^{\alpha\beta\gamma\delta}$$

$$= \eta_{00}\eta_{11}\eta_{22}\eta_{33}\epsilon^{0123} = -\epsilon^{0123},$$

$$\epsilon_{\mu\nu\rho\sigma} = -\epsilon^{\mu\nu\rho\sigma}. \tag{4.49}$$

We use the convention

$$\epsilon^{0123} = 1, \quad \epsilon^{0ijk} = \epsilon^{ijk}. \tag{4.50}$$

The Levi-Civita tensor $\epsilon^{\mu\nu\rho\sigma}$ transforms as

$$\epsilon'^{\mu\nu\rho\sigma} = \det(\Lambda)\,\Lambda^{\mu}{}_{\alpha}\,\Lambda^{\nu}{}_{\beta}\,\Lambda^{\rho}{}_{\gamma}\,\Lambda^{\sigma}{}_{\delta}\,\epsilon^{\alpha\beta\gamma\delta}. \tag{4.51}$$

It implies that

$$\epsilon'^{0123} = \det(\Lambda)\,\Lambda^{0}{}_{\alpha}\,\Lambda^{1}{}_{\beta}\,\Lambda^{2}{}_{\gamma}\,\Lambda^{3}{}_{\delta}\,\epsilon^{\alpha\beta\gamma\delta} = \det(\Lambda)^{2} = 1,$$

where

$$\det\Lambda = \Lambda^{0}{}_{\alpha}\,\Lambda^{1}{}_{\beta}\,\Lambda^{2}{}_{\gamma}\,\Lambda^{3}{}_{\delta}\,\epsilon^{\alpha\beta\gamma\delta}.$$

Now a tensor which transforms like

$$T'^{\mu\nu} = \det(\Lambda)\Lambda^{\mu}{}_{\rho}\Lambda^{\nu}{}_{\sigma}T^{\rho\sigma}, \tag{4.52}$$

is called a second rank contravariant pseudo tensor.

A contravariant axial vector or pseudo vector transforms as

$$A'^{\mu} = \det(\Lambda)\Lambda^{\mu}{}_{\rho}A^{\rho}. \tag{4.53}$$

If V^{μ} is a vector then

$$\begin{aligned}
V'_{\mu}A'_{\mu} &= V^{\alpha'}\eta_{\alpha\beta}A'^{\beta} \\
&= \Lambda^{\alpha}{}_{\rho}V^{\rho}_{\alpha\beta}\eta\det(\Lambda)\Lambda^{\beta}{}_{\sigma}A^{\sigma} \\
&= \det(\Lambda)\Lambda^{\alpha}{}_{\rho}\eta_{\alpha\beta}\Lambda^{\beta}{}_{\sigma}V^{\rho}A^{\sigma} \\
&= \det(\Lambda)\eta_{\rho\sigma}V^{\rho}A^{\sigma} = \det(\Lambda)V^{\rho}A_{\rho} \\
&= \det(\Lambda)V^{\mu}A_{\mu},
\end{aligned} \tag{4.54}$$

i.e., a pseudoscalar.

Chapter 5

FOUR-VELOCITY: MINKOWSKI FORCE

In this chapter we give a brief introduction to four-velocity, acceleration and Minkowski force.

5.1 Four-Velocity

Define four velocity U^μ

$$U^\mu = \frac{dx^\mu}{d\tau} = \gamma \frac{dx^\mu}{dt}, \tag{5.1}$$

where $d\tau$ is a proper time. In component form, Eq. (5.1) can be expressed as

$$U^i = \gamma \frac{dx^i}{dt} = \gamma(u)u^i,$$

$$U^0 = \gamma \frac{dx^0}{dt} = c\gamma(u), \tag{5.2}$$

$$U^\mu = (c\gamma(u), \gamma(u)\mathbf{u}).$$

Now

$$U^\mu U_\mu = \gamma^2(c^2 - \mathbf{u}^2) = \gamma^2 c^2 (1 - \frac{\mathbf{u}^2}{c^2}) = c^2, \tag{5.3}$$

i.e., U^2 is Lorentz invariant.

We can write Lorentz "boost" $\Lambda^\mu{}_\nu(u)$ given in Eq. (4.31) in a compact form in terms of $U^\mu \equiv (\gamma c, \gamma u^i)$ and $U_0^\mu \equiv (c, \mathbf{0})$

$$\Lambda^\mu{}_\nu(u) = \delta^\mu_\nu + \frac{(U + U_0)^\mu (U + U_0)_\nu}{c^2(\gamma + 1)} - \frac{2U^\mu U_{0\nu}}{\gamma c^2}.$$

Likewise, the 4-momentum $p^\mu = (p^0, \mathbf{p})$, with $p^0 = \frac{E}{c}$, $\mathbf{p}$ given in Eqs. (3.8) takes the form

$$p^\mu = mU^\mu. \tag{5.4}$$

5.2   Minkowski Force

Define Minkowski force in terms of the four-velocity U^μ as

$$f^\mu = \frac{dp^\mu}{d\tau} = m\frac{dU^\mu}{d\tau}. \tag{5.5}$$

Now

$$U^\mu f_\mu = mU^\mu \frac{dU_\mu}{d\tau} = \frac{1}{2}m\frac{d}{d\tau}\left(U^\mu U_\mu\right)$$

$$= \frac{1}{2}m\frac{d}{d\tau}\left(c^2\right) = 0. \tag{5.6}$$

In component form

$$U^0 f_0 - \mathbf{U} \cdot \mathbf{f} = c\gamma f_0 - \gamma^2 \mathbf{u} \cdot \mathbf{F} = 0 \tag{5.7}$$

where

$$\mathbf{f} = \frac{d\mathbf{p}}{d\tau} = \gamma\frac{d\mathbf{p}}{dt} = \gamma\mathbf{F} \tag{5.8}$$

and

$$\mathbf{F} = \frac{d\mathbf{p}}{dt} = m\frac{d}{dt}\left[\gamma\left(u\right)\mathbf{u}\right], \tag{5.9}$$

is the Newtonian force. Hence from Eqs. (5.7) and (5.8), one gets

$$f^0 = \gamma\frac{1}{c}\mathbf{u} \cdot \mathbf{F},$$

$$f^\mu = \gamma\left(\frac{1}{c}\mathbf{u} \cdot \mathbf{F}, \mathbf{F}\right). \tag{5.10}$$

Finally, the four-acceleration can be defined as

$$\alpha^\mu = \frac{dU^\mu}{d\tau} = \frac{d^2x^\mu}{d\tau^2}, \tag{5.11}$$

$$\alpha^i = \frac{dU^i}{d\tau} = \frac{dU^i}{dt}\frac{dt}{d\tau} = \gamma\frac{dU^i}{dt}$$

$$= \gamma\frac{d}{dt}\left(\gamma u^i\right), \tag{5.12}$$

$$\alpha^0 = \gamma\frac{d}{dt}\left(c\gamma\right), \tag{5.13}$$

α^μ being a four-vector, it transforms under Lorentz transformations

$$\boldsymbol{\alpha}' = \boldsymbol{\alpha} + \mathbf{v}[(\gamma(v) - 1)\frac{\mathbf{v}.\boldsymbol{\alpha}}{v^2} - \frac{\gamma(v)}{c}\alpha^0], \tag{5.14}$$

where

$$\boldsymbol{\alpha} = \gamma \frac{d}{dt} \left[\frac{\mathbf{u}}{\sqrt{1 - \frac{u^2}{c^2}}} \right]$$

$$= \gamma^2 \left[\mathbf{a} + \gamma^2 \mathbf{u} \frac{(\mathbf{u} \cdot \mathbf{a})}{c^2} \right], \tag{5.15}$$

$$\alpha^0 = \gamma^4 \frac{\mathbf{u} \cdot \mathbf{a}}{c}. \tag{5.16}$$

Let S^0 be a frame in which particle is instantaneously at rest, i.e., a frame moving with the particle $(\mathbf{u} = \mathbf{0})$. Thus in this frame

$$\alpha_0^\mu = (0, \mathbf{a}). \tag{5.17}$$

In a frame S relative to which frame S^0 is moving with velocity $\mathbf{u}$, Eq. (5.14) gives

$$\boldsymbol{\alpha} = \mathbf{a} + \mathbf{u} \left(\gamma - 1 \right) \frac{\mathbf{u} \cdot \mathbf{a}}{u^2}. \tag{5.18}$$

Let us apply the above equation to a particle which moves from rest at the origin in frame S along x-axis with acceleration g relative to its instantaneous rest frame (S^0). For this case

$$\alpha = g + (\gamma - 1)g = \gamma g. \tag{5.19}$$

Hence, the equation of motion takes the form

$$\gamma \frac{d}{dt} (\gamma u) = \gamma g,$$

giving

$$\frac{u}{\sqrt{1 - \frac{u^2}{c^2}}} = gt + \text{constant}. \tag{5.20}$$

With the condition $u = 0$ at $t = 0$, one has

$$u = \frac{dx}{dt} = \frac{gt}{\sqrt{1 + \frac{g^2 t^2}{c^2}}}. \tag{5.21}$$

Integration gives

$$x = \frac{c}{g} \left(c^2 + g^2 t^2 \right)^{1/2} + \text{constant}.$$

Taking $x = 0$ at $t = 0$ gives

$$x = \frac{c^2}{g} \left(\sqrt{1 + \frac{g^2 t^2}{c^2}} - 1 \right). \tag{5.22}$$

In the non-relativistic limit $\frac{g^2t^2}{c^2} \ll 1$, $\sqrt{1 + \frac{g^2t^2}{c^2}} \approx 1 + \frac{1}{2}\frac{g^2t^2}{c^2}$, we get

$$x = \frac{1}{2}gt^2$$

and

$$u = gt, \tag{5.23}$$

the well-known result in classical mechanics. The proper time τ is given by

$$d\tau = \sqrt{1 - \frac{u^2}{c^2}}\, dt$$

$$= \frac{dt}{\sqrt{1 + \frac{g^2t^2}{c^2}}}. \tag{5.24}$$

Integration gives

$$\tau = \frac{c}{g}\sinh^{-1}\frac{gt}{c}.$$

Thus

$$\frac{gt}{c} = \sinh\frac{g\tau}{c}. \tag{5.25}$$

Using Eq. (5.21), we obtain u/c in terms of proper time τ:

$$\frac{u}{c} = \tanh\frac{g\tau}{c}$$

$$\gamma(u) = \cosh\frac{g\tau}{c}. \tag{5.26}$$

The trajectory of the particle moving with a constant acceleration g in terms of proper time τ can be obtained as follows: From Eq. (5.26), one has

$$\frac{1}{c}\frac{dx}{d\tau} = \frac{\gamma}{c}\frac{dx}{dt} = \gamma\tanh\frac{g\tau}{c} = \sinh\frac{g\tau}{c}. \tag{5.27}$$

Integrating with boundary condition, at $\tau = 0$, $x = 0$, we get

$$x = \frac{c^2}{g}\left[\cosh\frac{g\tau}{c} - 1\right]. \tag{5.28}$$

Finally, the rapidity

$$\eta = \tanh^{-1}\frac{u}{c} = \frac{g\tau}{c} \tag{5.29}$$

$$x = \frac{c^2}{g}[\cosh\eta - 1]. \tag{5.30}$$

The last equation (5.30) gives the trajectory in terms of the rapidity.

5.3 Problems

5.1 For a Newtonian force

$$\mathbf{F} = \frac{d\mathbf{p}}{dt}, \ \mathbf{p} = \frac{m\mathbf{u}}{\sqrt{1 - \frac{u^2}{c^2}}}$$

$$E^2 = \mathbf{p}^2 c^2 + m^2 c^4 \,.$$

Show that:

(i)

$$\frac{dE}{dt} = \mathbf{u} \cdot \mathbf{F}.$$

(ii) The Lorentz transformations

$$x' = \gamma\left(x - vt\right), \ y' = y, \ z' = z, \ t' = \gamma\left(t - \frac{vx}{c^2}\right)$$

$$p'_x = \gamma\left(p_x - \frac{v}{c^2}E\right), \ \text{etc.}$$

Derive the relations

$$F'_x = \frac{F_x - \frac{v}{c^2}\mathbf{u} \cdot \mathbf{F}}{1 - \frac{v}{c^2}u_x}$$

$$F'_y = F_y \frac{1}{\gamma\left(1 - \frac{v}{c^2}u_x\right)}$$

$$F'_z = F_z \frac{1}{\gamma\left(1 - \frac{v}{c^2}u_x\right)}$$

if the particle is instantaneously at rest in S, i.e., $\mathbf{u} = \mathbf{0}$, then

$$\mathbf{F}'_\perp = \frac{1}{\gamma}\mathbf{F}_\perp, \ \mathbf{F}'_\parallel = \mathbf{F}_\parallel \,.$$

5.2 Using a fact that U^μ is a 4-vector and thus transforms, as

$$U'^\mu = \Lambda^\mu{}_\nu U^\nu$$

derive the relations:

(i)

$$\frac{\gamma\left(u'\right)}{\gamma\left(u\right)} = \gamma\left(v\right)\left(1 - \frac{\mathbf{u} \cdot \mathbf{v}}{c^2}\right).$$

(ii)

$$\mathbf{u}' = \frac{\gamma\left(u'\right)}{\gamma\left(u\right)}\left\{\mathbf{u} + \left(\gamma - 1\right)\frac{\mathbf{u} \cdot \mathbf{v}}{v^2}\mathbf{v} - \gamma\mathbf{v}\right\},$$

where
$$\gamma \equiv \gamma(v) = \frac{1}{\sqrt{1 - v^2/c^2}}.$$

Solutions:

$$U^\mu = \frac{dx^\mu}{d\tau} = \gamma(u)\frac{dx^\mu}{dt} = \gamma(u)u^\mu.$$

Thus
$$U^0 = c\gamma(u),$$
$$U^i = \gamma(u)\frac{dx^i}{dt} = \gamma(u)u^i.$$

Now
$$U'^\mu = \Lambda^\mu_{\ \nu}U^\nu$$

(i)
$$U'^0 = \Lambda^0_{\ \nu}U^\nu = \Lambda^0_{\ 0}U^0 + \Lambda^0_{\ i}U^i$$
$$= \gamma(v)c\gamma(u) - \gamma\frac{v^i}{c}\gamma(u)u^i$$
$$= c\gamma\gamma(u)\left(1 - \frac{\mathbf{u}\cdot\mathbf{v}}{c^2}\right).$$

Thus
$$c\gamma(u') = c\gamma\gamma(u)\left(1 - \frac{\mathbf{u}\cdot\mathbf{v}}{c^2}\right) \qquad (5.31)$$
$$\frac{\gamma(u')}{\gamma(u)} = \gamma\left(1 - \frac{\mathbf{u}\cdot\mathbf{v}}{c^2}\right).$$

(ii)
$$U'^i = \Lambda^i_{\ \nu}U^\nu = \Lambda^i_{\ 0}U^0 + \Lambda^i_{\ j}U^j$$
$$= \gamma(u)\left(-\gamma v^i + u^i + v^i\frac{\mathbf{u}\cdot\mathbf{v}}{v^2}(\gamma - 1)\right).$$

Thus
$$\gamma(u')u'^i = \gamma(u)\left(u^i + (\gamma - 1)\frac{\mathbf{u}\cdot\mathbf{v}}{v^2}v^i - \gamma v^i\right).$$

Therefore
$$\mathbf{u}' = \frac{\gamma(u)}{\gamma(u')}\left(\mathbf{u} + (\gamma - 1)\frac{\mathbf{u}\cdot\mathbf{v}}{v^2}\mathbf{v} - \gamma\mathbf{v}\right). \qquad (5.32)$$

From Eqs. (5.31) and (5.32)
$$\mathbf{u}' = \gamma^{-1}\left(1 - \frac{\mathbf{u}\cdot\mathbf{v}}{c^2}\right)\left(\mathbf{u} + (\gamma - 1)\frac{\mathbf{u}\cdot\mathbf{v}}{v^2}\mathbf{v} - \gamma\mathbf{v}\right).$$

Chapter 6

COVARIANT FORM OF
ELECTRODYNAMICS

In this chapter the covariant form of electrodynamics is formulated.

6.1 Electromagnetic Field Tensor

In 3-dimensional formulation, the electric and magnetic fields **E** and **B** in terms of scalar potential ϕ and vector potential **A** are given by

$$\mathbf{E} = -\nabla\phi - \frac{1}{c}\frac{\partial \mathbf{A}}{\partial t}, \tag{6.1}$$

$$\mathbf{B} = \nabla \times \mathbf{A}, \tag{6.2}$$

$$\mathbf{E} = (E_x, E_y, E_z)\,;E^i, \tag{6.3}$$

$$\mathbf{B} = (B_x, B_y, B_z)\,;B^i, \tag{6.4}$$

$$\mathbf{A} = (A_x, A_y, A_z)\,;A^i. \tag{6.5}$$

Now using

$$\partial^\mu = (\frac{1}{c}\frac{\partial}{\partial t}, -\nabla),$$

$$\partial_\mu = (\frac{1}{c}\frac{\partial}{\partial t}, \nabla), \tag{6.6}$$

the vector component of electric and magnetic fields $(E^i$ and $B^i)$ can be expressed as

$$E^i = \partial^i\phi - \partial^0 A^i, \tag{6.7}$$

$$B^i = \epsilon^{ijk}\partial_j A^k = -\epsilon^{ijk}\partial^j A^k, \tag{6.8}$$

so

$$\epsilon^{ijk}B^k = \partial^j A^i - \partial^i A^j. \tag{6.9}$$

In a 4-dimensional formalism $[A^0 = \phi]$

$$A^\mu = \left(A^0, \mathbf{A}\right) = \left(A^0, A^i\right).$$

Let us define the electromagnetic field tensor as

$$F^{\mu\nu} = \partial^\mu A^\nu - \partial^\nu A^\mu,$$

$$F_{\mu\nu} = \partial_\mu A_\nu - \partial_\nu A_\mu,$$

$$F^{\mu\nu} = -F^{\nu\mu}. \tag{6.10}$$

We note the cyclic property of derivative of tensor $F^{\mu\nu}$

$$\partial^\lambda F^{\mu\nu} = \partial^\lambda \left(\partial^\mu A^\nu - \partial^\nu A^\mu\right),$$

$$\partial^\mu F^{\nu\lambda} = \partial^\mu \left(\partial^\nu A^\lambda - \partial^\lambda A^\nu\right),$$

$$\partial^\nu F^{\lambda\mu} = \partial^\nu \left(\partial^\lambda A^\mu - \partial^\mu A^\lambda\right),$$

which then give

$$\partial^\lambda F^{\mu\nu} + \partial^\mu F^{\nu\lambda} + \partial^\nu F^{\lambda\mu} = 0. \tag{6.11}$$

Now from Eq. (6.10)

$$F^{0i} = \partial^0 A^i - \partial^i A^0 = -E^i,$$

$$F_{0i} = -F^{0i} \tag{6.12}$$

and

$$F^{ij} = \partial^i A^j - \partial^j A^i = -\epsilon^{ijk} B^k,$$

$$F^{12} = -B_z,$$

$$F^{23} = -B_x,$$

$$F^{31} = -B_y,$$

$$F^{ij} = F_{ij}. \tag{6.13}$$

Hence, in matrix form the electromagnetic field tensor becomes

$$F^{\mu\nu} = \begin{pmatrix} 0 & -E_x & -E_y & -E_z \\ E_x & 0 & -B_z & B_y \\ E_y & B_z & 0 & -B_x \\ E_z & -B_y & B_x & 0 \end{pmatrix} : (-\mathbf{E}, \mathbf{B}). \tag{6.14}$$

Define a dual tensor $\tilde{F}^{\mu\nu}$:

$$\tilde{F}^{\mu\nu} = -\frac{1}{2}\epsilon^{\mu\nu\rho\sigma} F_{\rho\sigma}, \tag{6.15}$$

which in component form gives

$$\tilde{F}^{0i} = -\frac{1}{2}\epsilon^{0ijk}F_{jk} = -\frac{1}{2}\epsilon^{ijk}F^{jk} \tag{6.16}$$

$$= B^i, \tag{6.17}$$

$$\tilde{F}^{ij} = -\frac{1}{2}\epsilon^{ijk0}F_{k0} - \frac{1}{2}\epsilon^{0ijk}F_{0k}$$

$$= \epsilon^{ijk}F^{0k}$$

$$= -\epsilon^{ijk}E^k. \tag{6.18}$$

Hence

$$\tilde{F}^{12} = -E^3 = -E_z,$$
$$\tilde{F}^{23} = -E^1 = -E_x,$$
$$\tilde{F}^{31} = -E^2 = -E_y, \tag{6.19}$$

which in matrix form is

$$\tilde{F}^{\mu\nu} = \begin{pmatrix} 0 & B_x & B_y & B_z \\ -B_x & 0 & -E_z & E_y \\ -B_y & E_z & 0 & -E_x \\ -B_z & -E_y & E_x & 0 \end{pmatrix} : (\mathbf{B}, \mathbf{E}). \tag{6.20}$$

6.1.1 Lorentz Force

In this section, the covariant form of Lorentz force is derived. The Lorentz force is given by (cf. Eq. (3.66))

$$F^i = e(E^i + \frac{1}{c}\epsilon^{ijk}v^j B^k)$$

$$= e[\partial^i A^0 - \partial^0 A^i - \frac{v^j}{c}\epsilon^{ijk}\epsilon^{klm}\partial^l A^m]$$

$$= e[\partial^i A^0 - \partial^0 A^i - \frac{v^j}{c}(\partial^i A^j - \partial^j A^i)]$$

$$= e[\partial^i A^0 - \partial^0 A^i - \frac{1}{c}\partial^i(\mathbf{v}\cdot A) + \frac{(v^j\partial^j)}{c}A^i]. \tag{6.21}$$

Now

$$\frac{d}{dx^0}A^i = \frac{\partial A^i}{\partial x^0} + \frac{dx^j}{dx^0}\frac{\partial A^i}{\partial x^j}$$

$$= \frac{\partial A^i}{\partial x^0} + \frac{v^j}{c}\partial_j A^i$$

$$= \frac{\partial A^i}{\partial x^0} - \frac{1}{c}v^j\partial^j A^i. \tag{6.22}$$

Thus

$$F^i = e[\partial^i A^0 - \frac{dA^i}{dx^0} - \frac{1}{c}\partial^i(\mathbf{v} \cdot \mathbf{A})]. \tag{6.23}$$

Now using the four-velocity

$$U^\mu = [c\gamma, \gamma\mathbf{v}], \tag{6.24}$$

$$U_\mu A^\mu = U_0 A^0 + U_i A^i,$$
$$= c\gamma A^0 - \gamma\mathbf{v} \cdot \mathbf{A}, \tag{6.25}$$

Eq. (6.23) is expressed in the form:

$$F^i = e[-\frac{1}{c}\frac{1}{\gamma}\frac{dA^i}{d\tau} + \frac{1}{c\gamma}\partial^i(U_\mu A^\mu)].$$

The corresponding Minkowski force is

$$f^i = \gamma F^i,$$

$$f^\mu = \frac{e}{c}\partial^\mu(U_\lambda A^\lambda) - \frac{dA^\mu}{d\tau}. \tag{6.26}$$

We can also write Lorentz force in terms of $F^{\mu\nu}$ on using Eqs. (6.12) and (6.13) as

$$F^i = e(-F^{0i} + \frac{v_j}{c}F^{ij})$$
$$= e(F^{i0} + F^{ij}\frac{v_j}{c}), \tag{6.27}$$

so that

$$F^i = \frac{e}{\gamma}\frac{1}{c}(F^{i0}U_0 + F^{ij}U_j),$$

$$\gamma F^i = \frac{e}{c}[F^{i\nu}U_\nu], \quad p_\nu = mU_\nu, \tag{6.28}$$

$$f^\mu = \frac{e}{mc}[F^{\mu\nu}p_\nu]$$

and

$$\frac{dp^\mu}{d\tau} = f^\mu = \frac{e}{mc}[F^{\mu\nu}p_\nu]. \tag{6.29}$$

This gives us the expression of Minkowski force in terms of the electromagnetic field tensor. Finally

$$F^\mu \equiv \frac{dp^\mu}{dt} = \frac{1}{\gamma}f^\mu = \frac{e}{\gamma mc}[F^{\mu\nu}p_\nu]. \tag{6.30}$$

6.2 Transformation of E and B Under Lorentz Transformation

The electromagnetic field tensor $F^{\mu\nu}$ transforms under Lorentz transformation as

$$F'^{\mu\nu} = \Lambda^\mu{}_\lambda \, \Lambda^\nu{}_\sigma \, F^{\lambda\sigma},$$
$$F'^{0i} = \Lambda^0{}_\lambda \, \Lambda^i{}_\sigma \, F^{\lambda\sigma},$$
$$F'^{ij} = \Lambda^i{}_\lambda \, \Lambda^j{}_\sigma \, F^{\lambda\sigma}. \tag{6.31}$$

Now (cf. Eq. (4.31))

$$\Lambda^0{}_0 = \gamma, \ \Lambda^0{}_i = -\gamma \frac{v^i}{c}, \ \Lambda^i{}_0 = -\gamma \frac{v^i}{c},$$

$$\Lambda^i{}_j = \delta^{ij} + \frac{v^i v^j}{v^2} \left(\gamma - 1 \right). \tag{6.32}$$

Thus

$$F'^{0i} = \Lambda^0{}_0 \, \Lambda^i{}_k \, F^{0k} + \Lambda^0{}_k \, \Lambda^i{}_0 \, F^{k0} + \Lambda^0{}_k \, \Lambda^i{}_j \, F^{kj}, \tag{6.33}$$

$$F'^{ij} = \left(\Lambda^i{}_0 \, \Lambda^j{}_k - \Lambda^i{}_k \, \Lambda^j{}_0 \right) F^{0k} + \Lambda^i{}_k \, \Lambda^j{}_n \, F^{kn}. \tag{6.34}$$

From Eq. (6.33), we have

$$-E'^i = -\gamma \left[\delta^{ik} + \frac{v^i v^k}{v^2} \left(\gamma - 1 \right) \right] E^k + \gamma^2 \frac{v^k v^i}{c^2} E^k - \gamma \frac{v^k}{c} \left[\delta^{ij} + \frac{v^i v^j}{v^2} \left(\gamma - 1 \right) \right] \epsilon^{jkl} B^l.$$

Hence noting that $\epsilon^{jkl} v^k v^j = 0$, the above equation takes the form

$$\mathbf{E'} = \gamma \left[\mathbf{E} + \frac{1 - \gamma}{\gamma} \frac{\mathbf{v} \cdot \mathbf{E}}{v^2} \mathbf{v} + \frac{1}{c} \left(\mathbf{v} \times \mathbf{B} \right) \right]. \tag{6.35}$$

Since

$$\tilde{F}^{0i} = B^i \ \text{ and } \ \tilde{F}^{ij} = -\epsilon^{ijk} E^k,$$

$$F^{0i} = -E^i \ \text{ and } \ F^{ij} = -\epsilon^{ijk} B^k,$$

and $\tilde{F}^{\mu\nu}$ transforms in the same way as $F^{\mu\nu}$, it follows

$$\mathbf{B'} = \gamma [\mathbf{B} + \frac{1 - \gamma}{\gamma} \frac{\mathbf{v} \cdot \mathbf{B}}{v^2} \mathbf{v} - \frac{1}{c} (\mathbf{v} \times \mathbf{E})]. \tag{6.36}$$

The inverse transformations are $(\mathbf{v} \to -\mathbf{v})$:

$$\mathbf{E} = \gamma \left[\mathbf{E'} + \frac{1 - \gamma}{\gamma} \frac{\mathbf{v} \cdot \mathbf{E'}}{v^2} \mathbf{v} - \frac{1}{c} \mathbf{v} \times \mathbf{B'} \right], \tag{6.37}$$

$$\mathbf{B} = \gamma \left[\mathbf{B'} + \frac{1 - \gamma}{\gamma} \frac{\mathbf{v} \cdot \mathbf{B'}}{v^2} \mathbf{v} + \frac{1}{c} \left(\mathbf{v} \times \mathbf{E'} \right) \right]. \tag{6.38}$$

In the non-relativistic limit, Eqs. (6.35) and (6.36) become

$$\mathbf{E}' = \mathbf{E} + \frac{1}{c}\mathbf{v} \times \mathbf{B} \tag{6.39}$$

$$\mathbf{B}' = \mathbf{B} - \frac{1}{c}\mathbf{v} \times \mathbf{E}.$$

6.3 Electromagnetic Field of a Moving Charge

As an application of Eqs. (6.35) and (6.36) let us calculate the electromagnetic field of a moving charge. For this purpose, consider a charged particle moving with velocity $\mathbf{v}$ with respect to an inertial frame S. Let S' be another inertial frame in which particle is at rest, i.e., a frame moving with the particle.

In S', the electromagnetic field due to a charged particle is given by

$$\mathbf{B}' = 0, \mathbf{E}' = \frac{e\mathbf{r}'}{r'^3}. \tag{6.40}$$

Using Eqs. (6.37) and (6.38), the electromagnetic fields of a moving charge (fields in S) are given by

$$\mathbf{E} = \gamma \frac{e}{r'^3}\left[\mathbf{r}' + \frac{1-\gamma}{\gamma}\frac{\mathbf{v}\cdot\mathbf{r}'}{v^2}\mathbf{v}\right],$$

$$\mathbf{B} = \gamma \frac{e}{r'^3}\frac{1}{c}\mathbf{v} \times \mathbf{r}', \tag{6.41}$$

where

$$\mathbf{r}' = \mathbf{r} - (1-\gamma)\,\mathbf{v}\frac{\mathbf{v}\cdot\mathbf{r}}{v^2}.$$

Thus

$$r' = \sqrt{\mathbf{r}'\cdot\mathbf{r}'} = r\left[1 + \gamma^2\frac{(\mathbf{v}\cdot\mathbf{r})^2}{c^2}\frac{1}{r^2}\right]^{1/2}. \tag{6.42}$$

Using Eqs. (6.42), we get from Eqs. (6.41)

$$\mathbf{E} = \gamma \frac{e}{r'^3}\mathbf{r} = \gamma \frac{e}{r^3}\mathbf{r}\frac{1}{\left[1 + \gamma^2\frac{(\mathbf{r}\cdot\mathbf{v})^2}{c^2 r^2}\right]^{3/2}}, \tag{6.43}$$

$$\mathbf{B} = \gamma \frac{e}{r'^3}\frac{1}{c}(\mathbf{v} \times \mathbf{r})$$

$$= \frac{1}{c}(\mathbf{v} \times \mathbf{E}). \tag{6.44}$$

In the non-relativistic limit, i.e., $v^2/c^2 << 1, \gamma \simeq 1$

$$\mathbf{E} = \frac{e}{r^3}\mathbf{r}, \tag{6.45}$$

$$\mathbf{B} = \frac{1}{c}\left(\mathbf{v} \times \mathbf{E}\right) = \frac{e}{r^3}\frac{1}{c}\left(\mathbf{v} \times \mathbf{r}\right). \tag{6.46}$$

6.4 Scalar and Vector Potential of a Moving Charge

In the frame S', the particle is instantaneously at rest, the scalar and vector potentials are given by

$$\mathbf{A}' = 0, \Phi' = \frac{e}{r'}, \tag{6.47}$$

$$A'^{\mu} \equiv (A'^0, \mathbf{A}') = (\Phi', 0).$$

Since A'^{μ} transforms as a 4-vector, hence vector and scalar potentials in the intertial frame S, relative to which the particle is moving with velocity $\mathbf{v}$ are given by

$$\mathbf{A} = \mathbf{A}' + \left[\mathbf{v}\frac{\mathbf{v} \cdot \mathbf{A}'}{v^2}\left(\gamma - 1\right) + \gamma\frac{\mathbf{v}}{c}\Phi'\right]$$

$$= \gamma\frac{\mathbf{v}}{c}\frac{e}{r'},$$

$$\Phi = \gamma\left[\Phi' + \mathbf{v} \cdot \mathbf{A}'\right] = \gamma\frac{e}{r'}, \tag{6.48}$$

where

$$r' = r\left[1 + \frac{(\mathbf{v} \cdot \mathbf{r})^2}{c^2}\frac{1}{r^2}\frac{1}{1 - \frac{v^2}{c^2}}\right]^{1/2}$$

$$= r\left[1 + \frac{v_r^2}{c^2}\frac{1}{1 - \frac{v_r^2 + v_t^2}{c^2}}\right]^{1/2}$$

$$= r\left[1 - \frac{v_t^2}{c^2}\right]^{1/2}\frac{1}{\left(1 - \frac{v^2}{c^2}\right)^{1/2}}$$

$$= r\gamma\sqrt{1 - \frac{v_t^2}{c^2}}. \tag{6.49}$$

Hence

$$\mathbf{A} = \frac{e\mathbf{v}/c}{r\sqrt{1 - \frac{v_t^2}{c^2}}},$$

$$\Phi = \frac{e}{r\sqrt{1 - \frac{v_t^2}{c^2}}}, \tag{6.50}$$

where v_t is the transverse velocity of the particle.

6.5 Covariant Form of Maxwell's Equations

The Maxwell's equations are

$$\nabla \cdot \mathbf{E} = 4\pi\rho, \tag{6.51}$$

$$\nabla \times \mathbf{B} = \frac{4\pi}{c}\mathbf{J} + \frac{1}{c}\frac{\partial \mathbf{E}}{\partial t}, \tag{6.52}$$

$$\nabla \cdot \mathbf{B} = 0, \tag{6.53}$$

$$\nabla \times \mathbf{E} = -\frac{1}{c}\frac{\partial \mathbf{B}}{\partial t}. \tag{6.54}$$

In addition one can write equation of continuity

$$\frac{\partial \rho}{\partial t} + \nabla \cdot \mathbf{J} = 0. \tag{6.55}$$

This equation can be obtained from Eqs. (6.51) and (6.52). Taking divergence of Eq. (6.52), we get

$$\nabla \cdot (\nabla \times \mathbf{B}) = \frac{4\pi}{c}\nabla \cdot \mathbf{J} + \frac{1}{c}\frac{\partial}{\partial t}(\nabla \cdot \mathbf{E}). \tag{6.56}$$

Making use of Eq. (6.51) one can easily check that we will get Eq. (6.55). In order to express Eq. (6.55) in covariant form, let us introduce a four-vector

$$J_\mu = (J_0, -\mathbf{J}) = (\rho c, -\mathbf{J}). \tag{6.57}$$

Using

$$\partial^\mu = \left(\frac{1}{c}\frac{\partial}{\partial t}, -\nabla\right). \tag{6.58}$$

The four-divergence of current J_μ gives

$$\partial^\mu J_\mu = \partial^0 J_0 + \partial^i J_i = \frac{\partial \rho}{\partial t} + \nabla \cdot \mathbf{J}.$$

Hence Eq. (6.55) can be written as

$$\partial^\mu J_\mu = 0.$$

Now

$$(\nabla \times \mathbf{B})^i = \epsilon^{ijk}\nabla^j B^k = \nabla^j(-F^{ij})$$
$$= -\partial_j F^{ij} = \partial_j F^{ji}, \tag{6.59}$$

$$\frac{1}{c}\left(\frac{\partial \mathbf{E}}{\partial t}\right)^i = \partial_0 E^i$$
$$= -\partial_0 F^{0i}, \tag{6.60}$$

$$\nabla \cdot \mathbf{E} = \partial_j E^j = -\partial_j F^{0j} = \partial_j F^{j0} = 4\pi\rho. \tag{6.61}$$

Therefore,

$$\left[(\nabla \times \mathbf{B}) - \frac{1}{c} \frac{\partial}{\partial t} \mathbf{E} \right]^i = \partial_j F^{ji} + \partial_0 F^{0i} = \frac{4\pi}{c} J^i. \tag{6.62}$$

Combining Eqs. (6.61) and (6.62)

$$\partial_j F^{ji} + \partial_0 F^{0i} + \partial_j F^{j0} = \frac{4\pi}{c} \left[J^i + J^0 \right]. \tag{6.63}$$

Thus the above equation in covariant form is written as

$$\partial_\mu F^{\mu\nu} = \frac{4\pi}{c} J^\nu. \tag{6.64}$$

Following the same procedure with Eq. (6.54)

$$(\nabla \times \mathbf{E})^i = \epsilon^{ijk} \nabla^j E^k = \nabla^j (\tilde{F}^{ij})$$
$$= \partial_j \tilde{F}^{ji}, \tag{6.65}$$

$$\left(\frac{1}{c} \frac{\partial \mathbf{B}}{\partial t} \right)^i = \partial_0 B^i$$
$$= \partial_0 \tilde{F}^{0i}, \tag{6.66}$$

$$0 = \nabla \cdot \mathbf{B} = \partial_j B^j$$
$$= \partial_j \tilde{F}^{0j} = -\partial_j \tilde{F}^{j0}. \tag{6.67}$$

Hence Eq. (6.67) can be written as

$$\left(\nabla \times \mathbf{E} + \frac{1}{c} \frac{\partial \mathbf{B}}{\partial t} \right)^i = \partial_j \tilde{F}^{ji} + \partial_0 \tilde{F}^{0i} = 0. \tag{6.68}$$

Combining Eqs. (6.67) and (6.68), one gets

$$\partial_j \tilde{F}^{ji} + \partial_0 \tilde{F}^{0i} + \partial_j \tilde{F}^{j0} = 0,$$
$$\partial_\nu \tilde{F}^{\nu\mu} = 0. \tag{6.69}$$

6.6 Energy-Momentum Tensor of an Electromagnetic Field

As is known the electromagnetic energy and momentum densities are given by

$$E = \frac{1}{8\pi} (\mathbf{E}^2 + \mathbf{B}^2), \quad p^i = \frac{1}{8\pi} (\mathbf{E} \times \mathbf{B})^i. \tag{6.70}$$

Now it is easy to check, using Eqs. (6.12) and (6.13) that

$$F_{\alpha\beta} F^{\alpha\beta} = -2\mathbf{E}^2 + 2\mathbf{B}^2, \tag{6.71}$$

$$F_{0\gamma} F^{0\gamma} = \mathbf{E}^2. \tag{6.72}$$

The energy momentum tensor is then given by

$$T^{\mu\nu} = \frac{1}{4\pi}(F^{\mu}{}_{\lambda}F^{\lambda\nu} + \frac{1}{4}g^{\mu\nu}F_{\alpha\beta}F^{\alpha\beta})$$

$$= \frac{1}{4\pi}(g^{\mu\rho}F_{\rho\lambda}F^{\lambda\nu} + \frac{1}{4}g^{\mu\nu}F_{\alpha\beta}F^{\alpha\beta}), \tag{6.73}$$

so that

$$T^{00} = \frac{1}{4\pi}[\mathbf{E}^2 + \frac{1}{4}(-2\mathbf{E}^2 + 2\mathbf{B}^2)]$$

$$= \frac{1}{8\pi}(\mathbf{E}^2 + \mathbf{B}^2), \tag{6.74}$$

$$T^{i0} = \frac{1}{4\pi}g^{i\rho}F_{\rho\lambda}F^{\lambda 0}$$

$$= \frac{1}{4\pi}g^{ij}F_{j\lambda}F^{\lambda 0} = \frac{1}{4\pi}g^{ij}F_{jk}F^{k0}$$

$$= \frac{\delta^i_j}{4\pi}(\epsilon^{jkl}B^l E^k)$$

$$= \frac{1}{4\pi}(\mathbf{E} \times \mathbf{B})^i, \tag{6.75}$$

$$T^{0i} = \frac{1}{4\pi}F_{0j}F^{ji} = \frac{1}{4\pi}(\mathbf{E} \times \mathbf{B})^i = T^{0i}. \tag{6.76}$$

Likewise

$$T^{ij} = \frac{1}{4\pi}\left[g^{ik}F_{k\lambda}F^{\lambda j} - \frac{1}{4}\delta^{ij}\left(F_{\alpha\beta}F^{\alpha\beta}\right)\right]$$

$$= -\frac{1}{4\pi}\left[F_{i0}F^{0j} + F_{in}F^{nj} + \frac{1}{4}\delta^{ij}\left(F_{\alpha\beta}F^{\alpha\beta}\right)\right]$$

$$= -\frac{1}{4\pi}\left[E^i E^j + B^i B^j - \frac{1}{2}\delta^{ij}\left(\mathbf{E}^2 + \mathbf{B}^2\right)\right].$$

Using Maxwell's equations in covariant form

$$\partial_\mu T^{\mu\nu} = \frac{1}{4\pi}[\partial^\rho(F_{\rho\lambda}F^{\lambda\nu}) + \frac{1}{4}\partial^\nu F_{\alpha\beta}F^{\alpha\beta}]$$

$$= \frac{1}{4\pi}[\partial^\rho F_{\rho\lambda}F^{\lambda\nu} + F_{\rho\lambda}\partial^\rho F^{\lambda\nu} + \frac{1}{2}F_{\alpha\beta}\partial^\nu F^{\alpha\beta}]$$

$$= \frac{1}{c}J_\lambda F^{\lambda\nu} + \frac{1}{8\pi}[F_{\alpha\beta}(\partial^\nu F^{\alpha\beta} + \partial^\alpha F^{\beta\nu}) + F_{\beta\alpha}\partial^\beta F^{\alpha\nu}]$$

$$= \frac{1}{c}J_\lambda F^{\lambda\nu} + \frac{1}{8\pi}F_{\alpha\beta}[\partial^\nu F^{\alpha\beta} + \partial^\alpha F^{\beta\nu} - \partial^\beta F^{\alpha\nu}]$$

$$= \frac{1}{c}J_\lambda F^{\lambda\nu} = -\frac{1}{c}F^{\nu\lambda}J_\lambda. \tag{6.77}$$

In order to get this relation, we have used Eq. (6.11).

Thus in the absence of sources, $T^{\mu\nu}$ is conserved. In the presence of sources we have to take into account the corresponding $T^{\mu\nu}$ for which as is well known

$$\partial_\mu T^{\mu\nu} = \partial^\nu \mathcal{L}, \tag{6.78}$$

where $\mathcal{L}$ is the Lagrangian density of the system of charges.

6.7 Problems

6.1 Show that

i) $\frac{1}{2}F^{\mu\nu}F_{\mu\nu} = (\mathbf{B}^2 - \mathbf{E}^2)$, ii) $\frac{1}{2}\widetilde{F}^{\mu\nu}F_{\mu\nu} = 2\mathbf{B}\cdot\mathbf{E}$.

6.2 The Lagrangian for the electromagnetic field is given by

$$L = -\frac{1}{4}F^{\mu\nu}F_{\mu\nu} - J^\mu A_\nu.$$

Using the Lagrange's equation of motion:

$$\frac{\partial L}{\partial A_\mu} - \partial_\nu \frac{\partial L}{\partial(\partial_\nu A_\mu)} = 0$$

derive the Maxwell's equations

$$\partial_\mu F^{\mu\nu} = J^\nu.$$

Show that addition of the term $\frac{1}{4}\widetilde{F}^{\mu\nu}F_{\mu\nu}$ in the Lagrangian does not change the Maxwell's equations.

6.3 Show that the equation

$$\varepsilon^{\mu\nu\rho\sigma}\partial_\nu F_{\rho\sigma} = 0$$

can be put in the form

$$\partial_\mu F_{\nu\rho} + \partial_\nu F_{\rho\mu} + \partial_\rho F_{\mu\nu} = 0.$$

6.4 Generalize the Larmor formula

$$P = \frac{2e^2}{3c^3}\overset{.}{v}{}^2 = \frac{2e^2}{3m^2c^3}\left(\frac{d\mathbf{p}}{dt}\cdot\frac{d\mathbf{p}}{dt}\right)$$

for the power radiated by a non-relativistic accelerated charge of mass m and momentum $\mathbf{p}$ to the relativistic charged particle.

i) Show that for linear motion along x-axis, the power radiated by an accelerated charge is given by

$$P = \frac{2}{3}\frac{e^2}{m^2c^3}\frac{1}{v^2}\left(\frac{dE}{dt}\right)^2.$$

Hence show that

$$\frac{\text{Power radiated}}{\text{Power supplied by external source}} \equiv \frac{P}{\left(\frac{dE}{dt}\right)} = \frac{2}{3}\frac{\frac{e^2}{mc^2}}{\beta mc^2}\left(\frac{dE}{dx}\right).$$

ii) Show that for a circular orbit of radius R, the power radiated per revolution is given by

$$\frac{PR}{2\pi c\beta} = \frac{4\pi}{3}\frac{e^2}{R}\gamma^4\beta^3.$$

Solution:

$$P = \frac{2e^2}{3c^3}\dot{v}^2 = \frac{2}{3}\frac{e^2}{m^2c^3}\left(\frac{d\vec{p}}{dt}\cdot\frac{d\vec{p}}{dt}\right).$$

For relativistic accelerated charge, the above equation is generalized to

$$P = \frac{2}{3}\frac{e^2}{m^2c^3}\left(-\frac{dp^\mu}{d\tau}\cdot\frac{dp_\mu}{d\tau}\right).$$

Now

$$\frac{dt}{d\tau} = \gamma,$$

$$p^0 = \frac{E}{c} = p_0.$$

That is

$$P = -\frac{2}{3}\frac{e^2}{m^2c^3}\gamma^2\left[\frac{dp^\mu}{dt}\frac{dp_\mu}{dt}\right]$$

$$= -\frac{2}{3}\frac{e^2}{m^2c^3}\gamma^2\left[\frac{1}{c^2}\left(\frac{dE}{dt}\right)^2 - \frac{d\vec{p}}{dt}\cdot\frac{d\vec{p}}{dt}\right].$$

(i)   For linear motion along x-axis

$$P = -\frac{2}{3}\frac{e^2}{m^2c^3}\gamma^2\left[\frac{1}{c^2}\left(\frac{dE}{dt}\right)^2 - \left(\frac{dp}{dt}\right)^2\right].$$

Now

$$v = \frac{c^2p}{E},$$

$$E^2 = c^2p^2 + m^2c^4,$$

$$2E\frac{dE}{dt} = 2c^2p\frac{dp}{dt}$$

$$= 2Ev\frac{dp}{dt}.$$

$$P = -\frac{2}{3}\frac{e^2}{m^2 c^3}\gamma^2[\frac{v^2}{c^2}-1](\frac{dp}{dt})^2 = \frac{2}{3}\frac{e^2}{m^2 c^3}\frac{1}{v^2}(\frac{dE}{dt})^2.$$

$$\frac{\text{Power radiated}}{\text{Power supplied by external sources}} = \frac{P}{\frac{dE}{dt}}$$

$$= \frac{2}{3}\frac{e^2}{m^2 c^3}\frac{1}{v^2}(\frac{dE}{dt})$$

$$= \frac{2}{3}\frac{e^2}{mc^2}\frac{1}{\beta mc^2}(\frac{dE}{dx}).$$

For circular orbit

$$\frac{dE}{dt} = 0, \frac{d\left(m\gamma c^2\right)}{dt} = 0,$$

$$\frac{d\vec{v}}{dt} = -\rho\omega^2\hat{\rho}.$$

$$\frac{d\vec{p}}{dt} = \frac{d}{dt}(m\gamma\vec{v})$$

$$= m\gamma\frac{d\vec{v}}{dt}.$$

Thus

$$P = \frac{2}{3}\frac{e^2}{m^2 c^3}\gamma^2[m^2\gamma^2\frac{d\vec{x}}{dt}\cdot\frac{d\vec{x}}{dt}]$$

$$= \frac{2}{3}\frac{e^2}{c^3}\gamma^4\rho^2\omega^4 = \frac{2}{3}\frac{e^2 c}{\rho^2}\gamma^4\beta^4.$$

Power radiated in one revolution

$$= PT = \frac{2}{3}\frac{e^2 c}{R^2}\gamma^4\beta^4(\frac{2\pi R}{c\beta})$$

$$= \frac{4\pi}{3}\frac{e^2}{R}\gamma^4\beta^3.$$

6.5 Consider a perfect fluid. Introduce a co-moving frame, i.e., a reference frame in which fluid is at rest. In this frame the fluid is completely determined by the energy-momentum tensor $T'^{ij} = p\delta^{ij}$, $T'^{00} = \rho c^2$, $T'^{i0} = 0$, $T'^{0i} = 0$. The co-moving frame is moving with velocity **v** with respect to fixed frame (Laboratory frame). The two frames are related by Lorentz transformation. Show that in the Laboratory frame, the energy-momentum tensor is given by

$$T^{00} = \gamma^2(\rho c^2 + \frac{v^2}{c^2}p), T^{0i} = T^{i0} = \gamma^2\frac{v^i}{c}(\rho c^2 + p),$$

$$T^{ij} = \frac{\gamma^2}{c^2}(\rho c^2 + p)v^i v^j + p\delta_{ij}.$$

Using the conservation of energy-momentum tensor

$$\partial_\mu T^{\mu\nu} = 0$$

derive the basic equations of hydrodynamics:

$$\frac{\partial \rho}{\partial t} + \nabla \cdot (\rho \mathbf{v}) = 0$$

$$\rho[\frac{\partial \mathbf{v}}{\partial t} + \mathbf{v} \cdot \nabla \mathbf{v}] = -\nabla \mathbf{p}$$

in the non-relativistic limit.

Solution: In the comoving frame, i.e., in the rest frame of fluid

$$\acute{T}^{00} = \rho c^2,$$
$$\acute{T}^{i0} = \acute{T}^{0i} = 0,$$
$$\acute{T}^{ij} = p\delta^{ij}.$$

The comoving frame is moving with velocity **v** with respect to fixed frame (lab frame). The two frames are related by Lorentz transformation. The lab frame is moving with respect to rest frame of fluid with $-\mathbf{v}$.

$$T^{\mu\nu} = \Lambda^\mu_\lambda \Lambda^\nu_\sigma \acute{T}^{\lambda\sigma},$$
$$T^{00} = \Lambda^0_\lambda \Lambda^0_\sigma \acute{T}^{\lambda\sigma}$$
$$= \left(\Lambda^0_0\right)^2 \acute{T}^{00} + \Lambda^0_i \Lambda^0_j \acute{T}^{ij}.$$

$$T^{i0} = \Lambda^i_\lambda \Lambda^0_\sigma \acute{T}^{\lambda\sigma}$$
$$= \Lambda^i_0 \Lambda^0_0 \acute{T}^{00} + \Lambda^i_0 \Lambda^0_j \acute{T}^{0j} + \Lambda^i_j \Lambda^0_0 \acute{T}^{ji} + \Lambda^i_j \Lambda^0_k \acute{T}^{jk},$$

$$T^{ij} = \Lambda^i_\lambda \Lambda^j_\sigma \acute{T}^{\lambda\sigma}$$
$$= \Lambda^i_0 \Lambda^j_\sigma \acute{T}^{0\sigma} + \Lambda^i_k \Lambda^j_\sigma \acute{T}^{k\sigma}$$
$$= \Lambda^i_0 \Lambda^j_\sigma \acute{T}^{00} + \Lambda^i_0 \Lambda^j_k \acute{T}^{0k} + \Lambda^i_k \Lambda^j_0 \acute{T}^{k0} + \Lambda^i_k \Lambda^j_n \acute{T}^{kn}.$$

Now using

$$\Lambda^0_0 = \gamma, \Lambda^i_0 = \gamma\frac{v^i}{c} = \Lambda^0_i,$$

$$\Lambda^{ij} = \delta^{ij} + \frac{v^i v^j}{v^2}(\gamma - 1),$$

we get

$$T^{00} = \gamma^2(\rho c^2 + \frac{v^2}{c^2}p),$$

$$T^{i0} = \frac{\gamma^2}{c}v^i(\rho c^2 + p) = T^{0i},$$

$$T^{ij} = \delta^{ij}p + \gamma^2\frac{v^i v^j}{c^2}(\rho c^2 + p).$$

The above equations can be written in covariant form:

$$T^{\mu\nu} = \frac{1}{c^2}(\rho c^2 + p)U^\mu U^\nu - \eta^{\mu\nu} p,$$

where

$$U^\mu U_\mu = \eta_{\mu\nu} U^\mu U^\nu = c^2,$$
$$U^\mu = (\gamma c, \gamma\vec{v}),$$
$$U^\mu = \frac{dx^\mu}{d\tau}$$
$$= \frac{dx^\mu}{dt}\frac{dt}{d\tau}$$
$$= v^\mu \gamma,$$
$$U^0 = \gamma c, \quad (\gamma^2 - 1) = \gamma^2 \frac{v^2}{c^2}.$$

Conservation of energy:

$$\partial_\mu T^{\mu\nu} = 0.$$

$\nu = 0$:

$$\partial_\mu T^{\mu 0} = \partial_0 T^{00} + \partial_i T^{i0} = 0.$$

In the non-relativistic limit ($v \ll c$)

$$\gamma^2 \to 1.$$

In the non-relativistic limit

$$T^{00} = \rho c^2,$$
$$T^{i0} = v^i c\rho,$$
$$T^{ij} = \delta^{ij} p + \rho v^i v^j.$$

Thus

$$\partial_\mu T^{\mu 0} = \frac{1}{c}\frac{\partial}{\partial t}(\rho c^2) + \partial_i v^i c\rho = 0.$$

Hence

$$\frac{\partial \rho}{\partial t} + \vec{\nabla} \cdot (\rho\vec{v}) = 0$$

is the equation of continuity.

$\nu = i$

$$\partial_\mu T^{\mu i} = \partial_0 T^{0i} + \partial_j T^{ji} = 0,$$
$$\partial_0 T^{0i} + \partial_j T^{ji} = \frac{c}{c}\frac{\partial}{\partial t}(\rho v^i) + \partial_j(\delta^{ij} p + \rho v^i v^j)$$
$$= \frac{\partial}{\partial t}(\rho v^i) + \partial_i p + \partial_j(\rho v^i v^j).$$

Using

$$\partial_j \left(v^i \rho v^j \right) = v^i \partial_j \left(\rho v^j \right) + \rho v^j \partial_j v^i$$
$$= v^i \nabla \cdot (\rho \mathbf{v}) + \rho (\mathbf{v} \cdot \nabla) v^i,$$

we have

$$\partial_0 T^{0i} + \partial_j T^{ji} = \frac{\partial}{\partial t} \left(\rho v^i \right) + \partial_i p + v^i \nabla \cdot (\rho \mathbf{v}) + \rho (\mathbf{v} \cdot \nabla) v^i = 0.$$

Thus

$$\rho \frac{\partial \mathbf{v}}{\partial t} + \mathbf{v} \frac{\partial \rho}{\partial t} + \nabla p + \mathbf{v} \nabla \cdot (\rho \mathbf{v}) + \rho (\mathbf{v} \cdot \nabla) \mathbf{v} = 0.$$

Now using the equation of continuity, we get

$$\rho [\frac{\partial \mathbf{v}}{\partial t} + (\mathbf{v} \cdot \boldsymbol{\nabla}) \mathbf{v}] = -\boldsymbol{\nabla} p.$$

Chapter 7

SPIN

7.1 Relativistic Equations of Motion for Spin in a Uniform External Electromagnetic Field

For a non-relativistic electron in a magnetic field $\mathbf{B}$, equation of motion for electron spin is given by

$$\frac{d\mathbf{s}}{dt} = \mu_{spin} \times \mathbf{B} = \frac{eg}{2mc} \mathbf{s} \times \mathbf{B}. \tag{7.1}$$

We want to write the equation of motion for the spin in a covariant form. For this purpose we define a four-vector (called the spin vector):

$$S^\mu = \left(S^0, \mathbf{S} \right),$$

such that in the rest frame, where $S^\mu = (0, \mathbf{s})$

$$S^\mu U_\mu = 0$$

$U_\mu = (\gamma c, -\gamma \mathbf{v})$ being four-velocity.

Thus in every frame

$$S^\mu U_\mu = 0 \text{ or } S^\mu p_\mu = 0,$$

$$S^0 = \frac{\mathbf{S} \cdot \mathbf{v}}{c}. \tag{7.2}$$

While S^0 vanishes by definition in any instantaneous rest-frame, $\frac{dS^0}{d\tau}$ need not. We note that we can rewrite Eq. (7.1) on using Eq. (6.13) as

$$\frac{dS^i}{d\tau} = \frac{eg}{2mc} \epsilon^{ijk} S^j B^k$$

$$= \frac{eg}{2mc} (-F^{ij} S^j)$$

$$= \frac{eg}{2mc} F^{ij} S_j.$$

71

A natural covariant generalization would be

$$\frac{dS^\mu}{d\tau} = \frac{eg}{2mc} F^{\mu\nu} S_\nu. \tag{7.3}$$

Now Eq. (7.2) implies

$$\frac{d}{d\tau}(S^\mu p_\mu) = 0,$$

$$S^\mu \frac{dp_\mu}{d\tau} + p_\mu \frac{dS^\mu}{d\tau} = 0. \tag{7.4}$$

We can guess the equation of motion for S^μ; it should be of the form

$$\frac{dS^\mu}{d\tau} = \frac{eg}{2mc} F^{\mu\nu} S_\nu + b_2 p^\mu \left(p_\alpha F^{\alpha\beta} S_\beta\right) + b_3 p^\mu \frac{dp^\lambda}{d\tau} S_\lambda. \tag{7.5}$$

The above equation follows from the following requirement; the right-hand side should be linear in S_ν and should be proportional to electromagnetic field tensor $F^{\mu\nu}$. The second and third terms should be quadratic in p, because under space reflection $\mathbf{p} \to -\mathbf{p}$, whereas $\mathbf{S} \to \mathbf{S}$.

Hence from Eqs. (7.4) and (7.5), we have (using $p^2 = mc^2$)

$$S^\mu \frac{dp_\mu}{d\tau} + \frac{eg}{2mc} p_\mu F^{\mu\nu} S_\nu + b_2 \left(m^2 c^2\right) p_\alpha F^{\alpha\beta} S_\beta + b_3 m^2 c^2 \frac{dp_\lambda}{d\tau} S_\lambda = 0, \tag{7.6}$$

giving the constraints

$$m^2 c^2 b_2 + \frac{eg}{2mc} = 0,$$

$$m^2 c^2 b_3 + 1 = 0. \tag{7.7}$$

From Eqs. (7.5) and (7.7), one gets

$$\frac{dS^\mu}{d\tau} = \frac{eg}{2mc} [F^{\mu\nu} S_\nu - \frac{1}{m^2 c^2} p^\mu (p_\alpha F^{\alpha\beta} S_\beta)] - \frac{1}{m^2 c^2} p^\mu \frac{dp^\lambda}{d\tau} S_\lambda. \tag{7.8}$$

Using the Lorentz equation of motion

$$\frac{dp^\mu}{d\tau} = \frac{e}{mc} F^{\mu\nu} p_\nu = -\frac{e}{mc} F^{\nu\mu} p_\nu, \tag{7.9}$$

we can write Eq. (7.8) as

$$\frac{dS^\mu}{d\tau} = \frac{eg}{2mc} F^{\mu\nu} S_\nu + (\frac{g}{2} - 1) \frac{1}{m^2 c^2} p^\mu \frac{dp^\lambda}{d\tau} S_\lambda, \tag{7.10}$$

or equivalently

$$\frac{dS^\mu}{d\tau} = \frac{eg}{2mc} F^{\mu\nu} S_\nu - (g - 2) \frac{e}{2mc} \frac{1}{m^2 c^2} p^\mu \left(p_\alpha F^{\alpha\beta} S_\beta\right). \tag{7.11}$$

Equation (7.11) is known as BMT equation of motion for the spin [V. Bargmann, I. Michel and V. L. Telegdi, Phys. Rev. Lett. 2 (1959) 435]. The above equation holds in any inertial frame. However in order to derive, Thomas precession and equation of helicity from Eq. (7.11), we consider equation of motion for the spin **s**.

Now

$$s^\mu = \Lambda^\mu{}_\nu(v)S^\nu, \tag{7.12}$$

which gives

$$0 = s^0 = \Lambda^0{}_0 S^0 + \Lambda^0{}_j S^j,$$
$$s^i = \Lambda^i{}_0 S^0 + \Lambda^i{}_j S^j. \tag{7.13}$$

Using Eq. (4.31) for $\Lambda^\mu{}_\nu$, we obtain

$$S^0 = \frac{v^j S^j}{c} = \frac{\mathbf{v}.\mathbf{S}}{c}, \tag{7.14}$$

$$s^i = -\gamma \frac{v^i}{c} S^0 + S^i + (\gamma - 1)v^i \frac{\mathbf{v}.\mathbf{S}}{v^2}. \tag{7.15}$$

Equations (7.14) and (7.15) give

$$s^i = S^i - \frac{v^i}{c}S^0 \frac{\gamma}{\gamma+1} = S^i - \frac{S^0}{mc}\frac{p^i}{\gamma+1}, \tag{7.16}$$

$$S^0 = \gamma\frac{\mathbf{v}.\mathbf{s}}{c} = \frac{\mathbf{p}.\mathbf{s}}{mc}. \tag{7.17}$$

From Lorentz Equation of motion (7.9):

$$mc\frac{d\gamma}{d\tau} = \frac{dp^0}{d\tau} = \frac{e}{mc}F^{0i}p_i = \frac{e}{mc}\mathbf{E}\cdot\mathbf{p}, \tag{7.18}$$

$$\frac{dp^i}{d\tau} = \frac{e}{mc}\left(F^{i0}p_0 + F^{ij}p_j\right)$$
$$= \frac{e}{mc}\left(E^i p_0 + \varepsilon^{ijk}B^k p^j\right)$$
$$= \frac{e}{mc}\left(m\gamma c E^i + (\mathbf{p}\times\mathbf{B})^i\right). \tag{7.19}$$

Note that $\frac{\mathbf{p}\cdot\mathbf{s}}{mc}$ is the helicity.

From Eq. (7.11), we have

$$\frac{dS^0}{d\tau} = \frac{1}{mc}\frac{d}{d\tau}(\mathbf{p}\cdot\mathbf{s}) = \frac{eg}{2mc}F^{0j}S_j - \frac{e(g-2)}{2m^3c^3}p^0\left(p_\alpha F^{\alpha\beta}S_\beta\right) \tag{7.20}$$

and from Eqs. (7.11) and (7.20), we obtain

$$\frac{ds^i}{d\tau} = \frac{dS^i}{d\tau} - \frac{1}{mc\,(\gamma+1)}\left(\frac{dS^0}{d\tau}p^i + S^0\frac{dp^i}{d\tau}\right) + \frac{S^0 p^i}{mc\,(\gamma+1)^2}\frac{d\gamma}{d\tau}$$

$$= \frac{eg}{2mc}\left(F^{i0}S_0 + F^{ij}S_j\right) - p^i\frac{e\,(g-2)}{2m^3c^3\,(\gamma+1)}\left(p_\alpha F^{\alpha\beta}S_\beta\right) - \frac{egp^i}{2m^2c^2\,(\gamma+1)}F^{0j}S_j$$

$$+ \frac{S^0 p^i}{mc\,(\gamma+1)^2}\frac{e}{m^2c^2}\left(\mathbf{E}\cdot\mathbf{p}\right) - \frac{S^0}{mc\,(\gamma+1)}\frac{e}{mc}\left(m\gamma cE^i + (\mathbf{p}\times\mathbf{B})^i\right).$$

$$(7.21)$$

Note that in writing the final form we have used Eqs. (7.19) and (7.18).

Now

$$F^{i0}S_0 = E^i\frac{\mathbf{p}\cdot\mathbf{s}}{mc}, \tag{7.22}$$

$$F^{ij}S_j = -\epsilon^{ijk}B^k\left(s_j + \frac{\mathbf{p}\cdot\mathbf{s}}{(\gamma+1)}\frac{p_j}{m^2c^2}\right)$$

$$= (\mathbf{s}\times\mathbf{B})^i + \frac{(\mathbf{p}\cdot\mathbf{s})\,(\mathbf{p}\times\mathbf{B})^i}{(\gamma+1)\,m^2c^2}, \tag{7.23}$$

$$F^{0j}S_j = (\mathbf{E}\cdot\mathbf{s}) + \frac{(\mathbf{p}\cdot\mathbf{E})\,(\mathbf{p}\cdot\mathbf{s})}{(\gamma+1)\,m^2c^2}, \tag{7.24}$$

$$p_\alpha F^{\alpha\beta}S_\beta = p_0 F^{0j}S_j + p_j F^{j0}S_0 + p_j F^{jk}S_k$$

$$= m\gamma c\,(\mathbf{E}\cdot\mathbf{s}) - \mathbf{p}\cdot(\mathbf{s}\times\mathbf{B}) - \frac{(\mathbf{p}\cdot\mathbf{E})\,(\mathbf{p}\cdot\mathbf{s})}{mc\,(\gamma+1)}. \tag{7.25}$$

Hence using Eqs. (7.22)–(7.25), we get from Eqs. (7.20) and (7.21)

$$\frac{d\mathbf{s}}{d\tau} = \frac{eg}{2mc}\,(\mathbf{s}\times\mathbf{B}) + \frac{e}{m^2c^2}\left(\frac{g}{2} - \frac{\gamma}{\gamma+1}\right)((\mathbf{p}\cdot\mathbf{s})\mathbf{E} - (\mathbf{E}\cdot\mathbf{s})\mathbf{p})$$

$$+ \frac{e}{m^3c^3\,(\gamma+1)}\left(\frac{g}{2} - 1\right)((\mathbf{p}\cdot\mathbf{s})(\mathbf{p}\times\mathbf{B}) - \mathbf{s}\cdot(\mathbf{p}\times\mathbf{B})\mathbf{p}). \tag{7.26}$$

Using the vector identities

$$\mathbf{s}\times(\mathbf{p}\times\mathbf{E}) = \mathbf{p}(\mathbf{s}\cdot\mathbf{E}) - \mathbf{E}(\mathbf{p}\cdot\mathbf{s}),$$

$$(\mathbf{p}\cdot\mathbf{s})(\mathbf{p}\times\mathbf{B}) - \mathbf{s}\cdot(\mathbf{p}\times\mathbf{B})\mathbf{p} = -(\mathbf{p}\cdot\mathbf{B})(\mathbf{s}\times\mathbf{p}) + \mathbf{p}^2(\mathbf{s}\times\mathbf{B}), \tag{7.27}$$

and $\mathbf{p}^2 = m^2c^2(\gamma^2 - 1)$, and $dt = \gamma d\tau$, we obtain from Eq. (7.26)

$$\frac{d\mathbf{s}}{dt} = -\frac{e}{m^2c^2}(\frac{g}{2} - \frac{\gamma}{\gamma+1})\frac{1}{\gamma}(\mathbf{s}\times(\mathbf{p}\times\mathbf{E})) + \frac{e}{mc}\left(\frac{g}{2} - 1 + \frac{1}{\gamma}\right)(\mathbf{s}\times\mathbf{B})$$

$$- \frac{e}{\gamma\,(\gamma+1)\,m^3c^3}\left(\frac{g}{2} - 1\right)((\mathbf{p}\cdot\mathbf{B})(\mathbf{s}\times\mathbf{p})). \tag{7.28}$$

The first term on the right-hand side of Eq. (7.28) gives the Thomas precession. The second term reduces to the precession of spin in uniform magnetic field in the non-relativistic limit. The last term demonstrates a remarkable property that in Dirac theory ($g = 2$) for electron and muon, this term vanishes. In order to see its significance, consider the equation of motion for the helicity ($\mathbf{s}.\mathbf{p}$). From Eq. (7.20)

$$\frac{d}{dt}(\mathbf{s} \cdot \mathbf{p}) = -\frac{e\,(g-2)}{2mc}\mathbf{s}\cdot(\mathbf{p} \times \mathbf{B}) + \frac{eg}{2\gamma}\left(\mathbf{s} \cdot \mathbf{E} + \frac{(\mathbf{p} \cdot \mathbf{E})(\mathbf{p} \cdot \mathbf{s})}{m^2c^2\,(\gamma + 1)}\right)$$
$$-\frac{e\,(g-2)}{2}\left(\gamma(\mathbf{E} \cdot \mathbf{s}) - \frac{(\mathbf{p} \cdot \mathbf{E})(\mathbf{p} \cdot \mathbf{s})}{m^2c^2\,(\gamma + 1)}\right). \tag{7.29}$$

We discuss the case $\mathbf{E} = \mathbf{0}$, as this is most interesting. For this case $\frac{d\gamma}{dt} = 0$, from Eq. (7.29):

$$\frac{d}{dt}(\mathbf{v} \cdot \mathbf{s}) = -\frac{e\,(g-2)}{2mc}\mathbf{s}\cdot(\mathbf{v} \times \mathbf{B}). \tag{7.30}$$

We have an important result: Since in Dirac theory $g = 2$, helicity is conserved. Any deviation would manifest itself in a slow variation of helicity in a uniform magnetic field. This is because deviation from $(g - 2)$ is due to radiative corrections to Dirac theory and are very small. The anomaly in $(g - 2)$ has been measured with high precision and agrees very well with the theory.

7.2   Problem

7.1 Consider an electron moving with velocity $\mathbf{v}$ relative to the nucleus. Viewed from the electron's rest frame, the nucleus is moving with velocity $-\mathbf{v}$. Thus the electron will experience a magnetic field

$$\mathbf{B}' = -\frac{1}{c}\mathbf{v} \times \mathbf{E}.$$

Due to the spin of the electron, the interaction Hamiltonian

$$H_{\text{spin}} = -\mu_{\text{spin}} \cdot \mathbf{B}'$$
$$= \frac{1}{c}\mu_{\text{spin}} \cdot (\mathbf{v} \times \mathbf{E})$$
$$= \frac{ge}{2m_ec^2}\mathbf{s} \cdot (\mathbf{v} \times \mathbf{E}).$$

Now

$$e\mathbf{E} = -\frac{\partial V}{\partial r}\frac{\mathbf{r}}{r} = -\frac{\mathbf{r}}{r}\frac{dV}{dr}.$$

Show that

$$H_{\text{spin}} = \frac{ge}{2m_e^2 c^2} \, (\mathbf{s} \cdot \mathbf{L}) \, \frac{1}{r} \frac{dV}{dr}. \qquad (7.31)$$

For hydrogen atom, the ground state and the first excited states are

$$n = 1, \ 1^1 s_{1/2}$$

$$n = 2, \ \begin{cases} 2^1 s_{1/2} \\ 2^1 p_{1/2}, \ 2^1 p_{3/2} \end{cases} \quad \text{degenerate}$$

The spin orbit coupling (cf. Eq. (7.31)) removes the degeneracy between $p_{1/2}$ and $p_{3/2}$ states but the splitting obtained from Eq. (7.31) is not in agreement with the experiment. For electron $g = 2$, however further relativistic corrections due to Thomas precession gives

$$H \, (\text{Thomas precession}) = -\frac{1}{2m_e c^2} \mathbf{s} \cdot (\mathbf{v} \times \mathbf{E})$$

$$= -\frac{1}{2m_e^2 c^2} \, (\mathbf{s} \cdot \mathbf{L}) \, \frac{1}{r} \frac{dV}{dr}$$

$$H = \frac{(g-1)\,e}{2m_e^2 c^2} \, (\mathbf{s} \cdot \mathbf{L}) \, \frac{1}{r} \frac{dV}{dr}$$

in agreement with the experiment.

Chapter 8

SPACE TIME GROUPS AND THEIR REPRESENTATIONS

8.1 Matrix Representation of Lorentz Transformation

The Lorentz transformation

$$x'^{\mu} = \Lambda^{\mu}_{\nu} \, x^{\nu} \tag{8.1}$$

can be written in matrix form as

$$x' = \Lambda \, x,$$
$$x = \Lambda^{T} \, x',$$
$$\Lambda^{T} \eta \Lambda = \eta. \tag{8.2}$$

Here x is a column matrix

$$x = \begin{pmatrix} x^0 \\ x^1 \\ x^2 \\ x^3 \end{pmatrix},$$

and η is a diagonal matrix with diagonal elements $(1, -1, -1, -1)$. Λ is a 4×4 Lorentz matrix

$$\Lambda = \begin{pmatrix} \Lambda^0_0 & \Lambda^0_1 & \Lambda^0_2 & \Lambda^0_3 \\ \Lambda^1_0 & \Lambda^1_1 & \Lambda^1_2 & \Lambda^1_3 \\ \Lambda^2_0 & \Lambda^2_1 & \Lambda^2_2 & \Lambda^2_3 \\ \Lambda^3_0 & \Lambda^3_1 & \Lambda^3_2 & \Lambda^3_3 \end{pmatrix}. \tag{8.3}$$

We can construct the matrix representation of Lorentz transformation as follows. For an infinitesimal transformation viz

$$\Lambda^{\mu}_{\ \nu} = \delta^{\mu}_{\ \nu} + \epsilon^{\mu}_{\ \nu}, \tag{8.4}$$

let us introduce an operator (matrix) $U(\Lambda)$

$$U(\Lambda) = 1 - \frac{1}{2}\epsilon_{\alpha\beta} A^{\alpha\beta}. \tag{8.5}$$

Then A gives the matrix representation of Lorentz transformation. From Eqs. (8.4) and (8.5), we have

$$-\frac{1}{2}\epsilon_{\alpha\beta} \left(A^{\alpha\beta}\right)^\mu{}_\nu = \epsilon^\mu{}_\nu$$
$$= \eta^{\mu\lambda}\epsilon_{\lambda\nu}$$
$$= \frac{1}{2}\eta^{\mu\lambda}\left(\epsilon_{\lambda\nu} - \epsilon_{\nu\lambda}\right)$$
$$= \frac{1}{2}\eta^{\mu\lambda}\left[\epsilon_{\alpha\beta}\delta^\alpha_\lambda\delta^\beta_\nu - \epsilon_{\alpha\beta}\delta^\alpha_\nu\delta^\beta_\lambda\right]$$
$$= \frac{1}{2}\epsilon_{\alpha\beta}\left[\eta^{\mu\lambda}\delta^\alpha_\lambda\delta^\beta_\nu - \eta^{\mu\lambda}\delta^\alpha_\nu\delta^\beta_\lambda\right],$$

which leads to

$$\left(A^{\alpha\beta}\right)^\mu_\nu = -\left[\eta^{\mu\alpha}\delta^\beta_\nu - \eta^{\mu\beta}\delta^\alpha_\nu\right],$$
$$A^{\alpha\beta} = -A^{\beta\alpha}. \tag{8.6}$$

Writing

$$A^{\alpha\beta} = +iM^{\alpha\beta},$$

Eq. (8.4) gives

$$U(\Lambda) = 1 - \frac{i}{2}\epsilon_{\alpha\beta} M^{\alpha\beta},$$

so that

$$\left(M^{\alpha\beta}\right)^\mu_\nu = i\left[\eta^{\mu\alpha}\delta^\beta_\nu - \eta^{\mu\beta}\delta^\alpha_\nu\right]. \tag{8.7}$$

Then six matrices

$$M^{ij} = M_{ij} = \left(M^{23}, M^{31}, M^{12}\right),$$
$$M^{0i} = -M_{0i} = \left(M^{01}, M^{02}, M^{03}\right),$$

are given by

$$\left(M^{12}\right)^{\mu}_{\nu} = i\left(\eta^{\mu 1}\delta^2_{\nu} - \eta^{\mu 2}\delta^1_{\nu}\right)$$

$$= \begin{pmatrix} 0 & 0 & 0 & 0 \\ 0 & 0 & -i & 0 \\ 0 & i & 0 & 0 \\ 0 & 0 & 0 & 0 \end{pmatrix} \equiv S^3 = S_z,$$

$$\left(M^{23}\right)^{\mu}_{\nu} = i\left(\eta^{\mu 2}\delta^3_{\nu} - \eta^{\mu 3}\delta^2_{\nu}\right)$$

$$= \begin{pmatrix} 0 & 0 & 0 & 0 \\ 0 & 0 & 0 & 0 \\ 0 & 0 & 0 & -i \\ 0 & 0 & i & 0 \end{pmatrix} \equiv S^1 = S_x,$$

$$\left(M^{31}\right)^{\mu}_{\nu} = i\left(\eta^{\mu 3}\delta^1_{\nu} - \eta^{\mu 1}\delta^3_{\nu}\right)$$

$$= \begin{pmatrix} 0 & 0 & 0 & 0 \\ 0 & 0 & 0 & i \\ 0 & 0 & 0 & 0 \\ 0 & -i & 0 & 0 \end{pmatrix} \equiv S^2 = S_y. \qquad (8.8)$$

$\mathbf{S} \equiv (S_x, S_y, S_z)$ gives the matrix representation of rotation subgroup of Lorentz group.

Now

$$\left(M^{01}\right)^{\mu}_{\nu} = i\left(\eta^{\mu 0}\delta^1_{\nu} - \eta^{\mu 1}\delta^0_{\nu}\right)$$

$$= i\begin{pmatrix} 0 & 1 & 0 & 0 \\ 1 & 0 & 0 & 0 \\ 0 & 0 & 0 & 0 \\ 0 & 0 & 0 & 0 \end{pmatrix} \equiv \mathcal{K}^1,$$

$$\left(M^{02}\right)^{\mu}_{\nu} = i\left(\eta^{\mu 0}\delta^2_{\nu} - \eta^{\mu 2}\delta^0_{\nu}\right)$$

$$= i\begin{pmatrix} 0 & 0 & 1 & 0 \\ 0 & 0 & 0 & 0 \\ 1 & 0 & 0 & 0 \\ 0 & 0 & 0 & 0 \end{pmatrix} = \mathcal{K}^2,$$

$$\left(M^{03}\right)^{\mu}_{\nu} = i\left(\eta^{\mu 0}\delta^3_{\nu} - \eta^{\mu 3}\delta^0_{\nu}\right)$$

$$= i\begin{pmatrix} 0 & 0 & 0 & 1 \\ 0 & 0 & 0 & 0 \\ 0 & 0 & 0 & 0 \\ 1 & 0 & 0 & 0 \end{pmatrix} = \mathcal{K}^3. \qquad (8.9)$$

$\mathcal{K} \equiv (\mathcal{K}_x, \mathcal{K}_y, \mathcal{K}_z)$ gives the matrix representation of Lorentz boost.

The matrices S_i, S_j, $\mathcal{K}_i$ and $\mathcal{K}_j$ satisfy the following commutation relations

$$[S_i, S_j] = i\epsilon_{ijk}S_k,$$

$$[S_i, \mathcal{K}_j] = i\epsilon_{ijk}\mathcal{K}_k,$$

$$[\mathcal{K}_i, \mathcal{K}_j] = -i\epsilon_{ijk}S_k. \tag{8.10}$$

We note that 4×4 matrices S's are Hermitian but matrices $\mathcal{K}$'s are anti-Hermitian. Thus the finite representation of Lorentz group is not unitary.

In the next section, we derive the commutation relations for the generators of Lorentz group from the transformation properties of fields under a group of transformation.

8.2   Invariance: Representations of a Group

Consider a group of transformations A:

$$x' = Ax. \tag{8.11}$$

At the same physical point, a scalar field $\phi(x)$ has the same value in primed and unprimed coordinate systems:

$$\phi'(x') = \phi(x),$$

thus

$$\phi'(x) = \phi(A^{-1}x). \tag{8.12}$$

For any other field having components ϕ_a, one has

$$\phi'_a(x) = S_{ab}\phi_b(A^{-1}x). \tag{8.13}$$

Corresponding to a group of transformations A, there is a unitary operator U:

$$\phi'_a(x) = U\phi_a(x)U^\dagger,$$

$$= S_{ab}\phi_b(A^{-1}x). \tag{8.14}$$

Here S are matrices and form a representation of group A.

For an infinitesimal transformation

$$x' = (1 + \epsilon)x$$

$$\phi_a\left(A^{-1}x\right) = \phi_a\left[(1 - \epsilon)x\right]$$

$$= \phi_a - i\epsilon\delta_{ab}d\left(x, \frac{\partial}{\partial x}\right)\phi_b(x). \tag{8.15}$$

The corresponding unitarity operator is

$$U = 1 - i\epsilon F, \quad F^\dagger = F, \tag{8.16}$$

while

$$S_{ab} = (1 - i\epsilon T)_{ab} = \delta_{ab} - i\epsilon T_{ab}. \tag{8.17}$$

F is the generator of transformation and S or T are matrices operating on components of ϕ. U and F are operators in the Hilbert space of states and d's are differential operators. From Eq. (8.14), using Eqs. (8.16) and (8.17), we have

$$\begin{aligned}
\phi_a'(x) &= U\phi_a(x)U^\dagger \\
&= (1 - i\epsilon F)\,\phi_a(x)\,(1 + i\epsilon F) \\
&= \phi_a(x) + i\epsilon\,[\phi_a(x), F] \\
&= S_{ab}\phi_b(A^{-1}x) \\
&= (\delta_{ab} - i\epsilon T_{ab})\left[\phi_b(x) - i\epsilon\delta_{bc}d\left(x, \frac{\partial}{\partial x}\right)\phi_c\right] \\
&= \phi_a(x) - i\epsilon\delta_{ac}d\left(x, \frac{\partial}{\partial x}\right)\phi_c(x) - i\epsilon T_{ab}\phi_b(x). \tag{8.18}
\end{aligned}$$

This gives

$$[\phi_a(x), F] = \left[-T_{ab} - \delta_{ab}d\left(x, \frac{\partial}{\partial x}\right)\right]\phi_b(x)$$

or

$$[\phi(x), F] = [-T - d]\,\phi(x). \tag{8.19}$$

Taking the Hermitian conjugate, we get

$$\left[F, \phi^\dagger(x)\right] = \phi^\dagger(x)\left[-T^\dagger + d^\dagger\right]. \tag{8.20}$$

For successive transformations:

$$[F_1, [F_2, \phi]] = (T_1 + d_1)(T_2 + d_2)\,\phi, \tag{8.21}$$

$$[F_2, [F_1, \phi]] = (T_2 + d_2)(T_1 + d_1)\,\phi. \tag{8.22}$$

Subtracting Eq. (8.21) from (8.22), we obtain

$$[F_1, [F_2, \phi]] + [F_2, [\phi, F_1]] = ([T_1, T_2] + [d_1, d_2])\,\phi, \tag{8.23}$$

since T's and d's commute with each other. On using the Jacobi identity

$$[F_1, [F_2, \phi]] + [F_2, [\phi, F_1]] + [\phi, [F_1, F_2]] = 0, \tag{8.24}$$

we obtain from Eqs. (8.23) and (8.24)

$$[[F_1, F_2], \phi] = ([T_1, T_2] + [d_1, d_2])\,\phi. \tag{8.25}$$

Consistency requires that F's, T's and d's obey the same commutation relations. For any group of transformations the differential operators d's can be constructed and their commutation relations are worked out explicitly. These then determine the commutation relations for F's and T's.

8.3   Poincaré Group and its Representations

Consider the transformation

$$x^\mu \to x'^\mu = \Lambda^\mu{}_\nu x^\nu + a^\mu. \tag{8.26}$$

To find the generators, consider the infinitesimal transformation

$$
\begin{aligned}
x'^\mu &= x^\mu + \epsilon^\mu{}_\nu x^\nu + a^\mu \\
&= (\eta^{\mu\nu} + \epsilon^{\mu\nu}) x_\nu + a^\mu, \tag{8.27}
\end{aligned}
$$

where $\epsilon^\mu{}_\nu$ and a^μ are infinitesimal. Correspondingly the unitarity operator is

$$U_{a,\Lambda} = 1 + \frac{i}{2}\epsilon_{\mu\nu}M^{\mu\nu} - i a_\mu P^\mu. \tag{8.28}$$

We obtain from Eq. (8.18)

$$
\begin{aligned}
&\phi(x) - i\left[-\frac{1}{2}\epsilon_{\mu\nu}M^{\mu\nu} + a_\mu P^\mu, \phi(x)\right] \\
&= S\phi(x_\mu - \epsilon^{\mu\nu}x_\nu - a^\mu) \\
&= S\left[\phi(x) - \epsilon^{\mu\nu}x_\nu\partial_\mu\phi(x) - a^\mu\partial_\mu\phi(x)\right] \\
&= S\left[\phi(x) - \frac{1}{2}\epsilon^{\mu\nu}(x_\nu\partial_\mu - x_\mu\partial_\nu)\phi(x) - a^\mu\partial_\mu\phi(x)\right] \\
&= \left[1 - \frac{i}{2}\epsilon_{\mu\nu}S^{\mu\nu} - \frac{i}{2}\epsilon_{\mu\nu}L^{\mu\nu} - a_\mu\partial^\mu\right]\phi(x).
\end{aligned}
$$

Thus we obtain

$$\left[a_\mu P^\mu - \frac{1}{2}\epsilon_{\mu\nu}M^{\mu\nu}, \phi(x)\right] = \left[-i a_\mu\partial^\mu + \frac{1}{2}\epsilon_{\mu\nu}J^{\mu\nu}\right]\phi(x), \tag{8.29}$$

where

$$P_\mu = -i\partial_\mu$$

and

$$J^{\mu\nu} = S^{\mu\nu} + L^{\mu\nu},$$

$$L^{\mu\nu} = i(x^\mu\partial^\nu - x^\nu\partial^\mu). \tag{8.30}$$

The exponentiation gives

$$U_{a,\Lambda} = \exp\left[-i a_\mu P^\mu + \frac{i}{2}\epsilon_{\mu\nu}M^{\mu\nu}\right]. \tag{8.31}$$

There are 10 generators of the Poincaré group $P^\mu, M^{\mu\nu}$. As mentioned previously these generators should satisfy the same commutation relations which the corresponding differential operators $-i\partial^\mu$ and $L^{\mu\nu}$ do, the latter can be easily calculated. Thus

$$[P^\mu, P^\nu] \to (-i)^2 [\partial^\mu, \partial^\nu] = 0, \qquad (8.32)$$

$$\begin{aligned}
[M^{\mu\nu}, P^\sigma] &\to (-i) [L^{\mu\nu}, \partial^\sigma] \\
&= [(x^\mu \partial^\nu - x^\nu \partial^\mu), \partial^\sigma] \\
&= (\partial^\sigma x^\nu) \partial^\mu - (\partial^\sigma x^\mu) \partial^\nu \\
&= \frac{\partial x^\nu}{\partial x_\sigma} \partial^\mu - \frac{\partial x^\mu}{\partial x_\sigma} \partial^\nu \\
&= \eta^{\nu\sigma} \partial^\mu - \eta^{\mu\sigma} \partial^\nu \\
&\to i (\eta^{\nu\sigma} P^\mu - \eta^{\mu\sigma} P^\nu). \qquad (8.33)
\end{aligned}$$

$$\begin{aligned}
[M^{\mu\nu}, M^{\rho\sigma}] &\to [L^{\mu\nu}, L^{\rho\sigma}] \\
&= i \left(\begin{array}{c} \eta^{\mu\sigma} L^{\nu\rho} - \eta^{\nu\sigma} L^{\mu\rho} \\ +\eta^{\nu\rho} L^{\mu\sigma} - \eta^{\mu\rho} L^{\nu\sigma} \end{array} \right) \\
&\to i \left(\begin{array}{c} \eta^{\mu\sigma} M^{\nu\rho} - \eta^{\nu\sigma} M^{\mu\rho} \\ +\eta^{\nu\rho} M^{\mu\sigma} - \eta^{\mu\rho} M^{\nu\sigma} \end{array} \right). \qquad (8.34)
\end{aligned}$$

For a spinor field

$$S^{\mu\nu} = \frac{1}{2} \Sigma^{\mu\nu} = \frac{i}{2} [\gamma^\mu, \gamma^\nu], \qquad (8.35)$$

where γ's are Dirac matrices which anticommute with each other

$$\{\gamma^\mu, \gamma^\nu\} = \gamma^\mu \gamma^\nu + \gamma^\nu \gamma^\mu = 2\eta^{\mu\nu}. \qquad (8.36)$$

To gain more insight on these commutation relations, we separate the spatial and time components

$$P^\mu = (P^0, P^m),$$

and define

$$\begin{aligned}
\epsilon^{123} &= 1, \ \epsilon_{123} = -1, \\
M^{ij} &= -\varepsilon^{ijk} J_k = \varepsilon^{ijk} J^k, \\
M_{ij} &= -\varepsilon_{ijk} J^k = \varepsilon_{ijk} J_k,
\end{aligned} \qquad (8.37)$$

so that

$$M^{12} = -J_3 = J^3, \ M_{12} = J^3, \ \text{etc.}$$

From Eq. (8.34), one can easily write down the commutation relations for J's

$$\left[J^i, J^j\right] = i\epsilon^{ijk}J^k.$$

Thus the generators J^i satisfy the commutation relations of the rotation group O_3. Hence out of the 6 generators of Lorentz group, three generators M^{ij} belong to the subgroup O_3. The other three generators M^{0i} give the Lorentz boosts. Define

$$M^{0i} = K^i. \tag{8.38}$$

Then from Eq. (8.34), we get the commutation relations for K's:

$$\left[K^j, J^l\right] = i\epsilon^{jlm}K^m,$$
$$\left[K^i, K^j\right] = \left[M^{0i}, M^{0j}\right] = -i\eta^{00}M^{ij} = -i\epsilon^{ijk}J^k. \tag{8.39}$$

Note that the minus sign in Eq. (8.39) is the manifestation of the non-compactness of the Lorentz group. Note that the matrix representation of K's are non-Hermitian matrices. The negative sign originates from a Minkowski matrix, $\eta^{00} = 1$, $\eta^{ij} = -\delta_{ij}$. As a consequence the irreducible representations are radically different in nature from those of the rotation group.

In order to find the transformation matrix for the fundamental spinors $\left(\frac{1}{2}, 0\right)$ and $\left(0, \frac{1}{2}\right)$, we proceed as follows. The mathematical designation for proper homogeneous Lorentz group $L_+^\uparrow$ is $SO(3,1)$ which refers to the fact that signature of the metric $\eta^{\mu\nu}$ is $1, -1, -1, -1$.

Now $SO(3)$ is isomorphic with $SU(2)$. There is a natural correspondence between $L_+^\uparrow$ and $SL(2,C)$ where $SL(2,C)$ is special linear group in 2-dimensional complex space. We associate each space-time point x^μ with a 2×2 Hermitian matrix

$$x^\mu \to X = \sigma_\mu x^\mu, \tag{8.40}$$

where

$$\sigma_\mu = (\sigma_0, \sigma_1, \sigma_2, \sigma_3) \tag{8.41}$$
$$= (\sigma_0, -\sigma^i), \, i = 1, 2, 3.$$

σ_0 being a unit 2×2 matrix and σ^i are the usual Pauli matrices. Thus X is a 2×2 Hermitian matrix. It is easy to see that

$$\det X = \left(x^0\right)^2 - \left(x^1\right)^2 - \left(x^2\right)^2 - \left(x^3\right)^2$$
$$= \left(x^0\right)^2 - |\mathbf{x}|^2. \tag{8.42}$$

There is one-to-one correspondence between

$$x'^{\mu} = \Lambda^{\mu}{}_{\nu}x^{\nu} \text{ and } X' = AXA^{\dagger}, \tag{8.43}$$

where A is a 2×2 complex matrix $\begin{pmatrix} \alpha & \beta \\ \gamma & \delta \end{pmatrix}$. Now

$$\Lambda^{\mu}{}_{\nu} = \frac{1}{2}Tr\left(\sigma^{\mu}A\sigma_{\nu}A^{\dagger}\right). \tag{8.44}$$

This can be seen as follows: From Eqs. (8.40) and (8.43)

$$\sigma_{\mu}x'^{\mu} = A\sigma_{\nu}x^{\nu}A^{\dagger},$$

$$\sigma^{\lambda}\sigma_{\mu}\Lambda^{\mu}{}_{\nu}x^{\nu} = \sigma^{\lambda}A\sigma_{\nu}A^{\dagger}x^{\nu}.$$

Taking trace

$$2\delta^{\lambda}_{\mu}\Lambda^{\mu}{}_{\nu}x^{\nu} = Tr\left[\sigma^{\lambda}A\sigma_{\nu}A^{\dagger}\right]x^{\nu}$$

giving Eq. (8.44). Under Lorentz transformation Λ

$$\det X = x_0^2 - |\mathbf{x}|^2$$

is invariant. This implies

$$\det X = \det X' = \det A \cdot \det X \cdot \det A^{\dagger}.$$

Therefore

$$\det A \cdot \det A^{\dagger} = (\det A)^2 = 1$$

implying

$$\det A = \pm 1. \tag{8.45}$$

Since for $L_+^{\uparrow}$, $\det \Lambda = +1$, it follows that $\det A = 1$. The matrices A form a group under $SL(2, C)$. Hence we establish a natural correspondence between elements of Lorentz group $L_+^{\uparrow}$ and $SL(2, C)$. In this connection we note that Λ has six real parameters. A with $\det A = 1$ has also six real parameters. To each matrix A, there corresponds a Lorentz transformation Λ and vice versa. To a given Lorentz transformation Λ there corresponds two matrices A and $-A$. Recall that spinor representations are 2-values, $R(2\pi) = -1 \neq 1$. In particular two matrices $+I, -I$ in $SL(2, C)$ correspond to the identity element of $L_+^{\uparrow}$.

Now

$$\det A = \begin{vmatrix} A_1^1 & A_2^1 \\ A_1^2 & A_2^2 \end{vmatrix} = A_1^1 A_2^2 - A_1^2 A_2^1.$$

We note from Eq. (8.44) that $\Lambda^\mu{}_\nu$ remains the same if we replace A by A^* and $A^\dagger$ by $(A^*)^\dagger$. Thus for a given Lorentz matrix Λ, there are in general four matrices

$$A, -A, A^*, -A^*.$$

Note that A and A^* are not equivalent. This is because these matrices are in general not unitary so that one cannot find a similarity transformation relating A to A^*. Therefore A and A^* contribute two non-equivalent representation of $L_+^\uparrow$, acting on two different 2-dimensional vector spaces. Thus we have two non-equivalent bases, ξ, ξ^* which are called contravariant and covariant spinors. Spinor representations are called fundamental representations. They transform as

$$\xi^\alpha \longrightarrow \xi'^\alpha = A^\alpha_\beta \xi^\beta, \ \alpha, \beta = 1, 2$$

$$\xi' = A\xi \tag{8.46}$$

where

$$\xi = \begin{pmatrix} \xi^1 \\ \xi^2 \end{pmatrix}. \tag{8.47}$$

Correspondingly we write

$$\xi^* = \begin{pmatrix} \xi^{\dot{1}} \\ \xi^{\dot{2}} \end{pmatrix} \tag{8.48}$$

with the transformation law

$$\xi'^{\dot\alpha} = A^{*\dot\alpha}{}_{\dot\beta} \xi^{\dot\beta}. \tag{8.49}$$

Now $\frac{1}{2}\sigma^i$ and $-\frac{i}{2}\sigma^i$ provide 2×2 matrix representations for the generators J^i, K^i. For a spinor representation

$$J^i \to -\frac{1}{2}\sigma^i, \ K^i \to \frac{i}{2}\sigma^i. \tag{8.50}$$

Thus comparing with the unitary operator $U = e^{-i\omega \cdot \mathbf{J} - i\varsigma \cdot \mathbf{K}}$ associated with $L_+^\uparrow$

$$A = e^{\frac{1}{2}\left(i\theta \mathbf{n} \cdot \sigma + \varsigma \frac{\mathbf{v}}{v} \cdot \sigma\right)}, \tag{8.51}$$

where we have put $\varsigma = \varsigma \frac{\mathbf{v}}{v}$, $\omega = \theta \mathbf{n}$. For the classification of irreducible representation of $L_+^\uparrow$, it is useful to introduce Hermitian combinations

$$M^i = \frac{1}{2}\left(J^i + iK^i\right),$$

$$N^i = \frac{1}{2}\left(J^i - iK^i\right), \tag{8.52}$$

$$\left[M^i, M^j\right] = i\epsilon^{ijk} M^k,$$
$$\left[N^i, N^j\right] = i\epsilon^{ijk} N^k, \tag{8.53}$$
$$\left[N^i, M^j\right] = 0.$$

This algebra is identical to the Lie algebra of group $SU(2)_M \otimes SU(2)_N$ with two Casimir operators

$$M^2 = M^i M^i,$$
$$N^2 = N^i N^i, \tag{8.54}$$
$$\left[M^2, M^i\right] = 0, \left[N^2, N^i\right] = 0.$$

Since in Eq. (8.52) M^i and N^i are not linear combinations of basic elements, the commutators in Eqs. (8.47), (8.53) do not define the (real) Lie algebra of $L_+^\uparrow$. However, due to analogy with angular momentum, it is convenient to use the eigenvalues of M^2, M^3, N^2, N^3, which are respectively $u(u+1), v(v+1)$ $u, v = 0, 1/2, 1, 3/2, \ldots$. to label the elements of irreducible representation. This provides the basis $|kl\rangle$, where

$$|kl\rangle = \{|u, k\rangle |v, l\rangle\}$$
$$k = -u, \ldots, u, \ l = -v, \ldots, v.$$

Suppressing u, v on the basis vectors of the product space

$$J^3 |kl\rangle = \left(M^3 + N^3\right) |kl\rangle$$
$$= M^3 |u, k\rangle |v, l\rangle + |u, k\rangle N^3 |v, l\rangle$$
$$= (k + l) |k, l\rangle,$$
$$J^\pm |kl\rangle = |k \pm 1, l\rangle \left[u(u+1) - k(k \pm 1)\right]^{1/2}$$
$$+ |k, l \pm 1\rangle \left[v(v+1) - l(l \pm 1)\right]^{1/2}, \tag{8.55}$$
$$K^3 |k, l\rangle = |k, l\rangle i(l - k),$$
$$K^\pm |k, l\rangle = |k, l \pm 1\rangle i \left[v(v+1) - l(l \pm 1)\right]^{1/2}$$
$$- |k \pm 1, l\rangle \left[u(u+1) - k(k \pm 1)\right]^{1/2}. \tag{8.56}$$

We see that in the (u, v) representation, matrix representation of K^3 is diagonal and imaginary. This implies that finite dimensional representation of the Lorentz groups are non-unitary. We can label the representation of Lorentz group as follows:

u, v	Representation
$0, 0$	Lorentz scalar
$(1/2, 0)$	2-component Lorentz spinor of the first kind
$(0, 1/2)$	2-component Lorentz spinor of the second kind
$(1/2, 1/2)$	Lorentz four-vector

From Eq. (8.48), it follows that under space reflection $M \longleftrightarrow N$ and thus $(1/2, 0) \longleftrightarrow (0, 1/2)$. Hence if u left-handed spinor belongs to representation $(1/2, 0)$, the right-handed spinor belongs to representation $(0, 1/2)$.

Now for fundamental spinor representation

$$J^i = -\frac{1}{2}\sigma^i, \ K^i = \frac{i}{2}\sigma^i, \tag{8.57}$$

$$M^i = -\frac{\sigma^i}{2}, \ N^i = 0,$$

$$A^* = e^{\frac{1}{2}\left(-i\theta\sigma^* \cdot \mathbf{n} + \varsigma\sigma^* \cdot \frac{\mathbf{v}}{v}\right)}$$

so that correspondingly

$$J^i \to \frac{1}{2}\sigma^{*i}, \ K^i \to \frac{i}{2}\sigma^{*i}. \tag{8.58}$$

Since

$$\sigma^2 \sigma^{*i} \sigma^2 = -\sigma^i, \tag{8.59}$$

$$\sigma^2 A^* \sigma^2 = e^{\frac{1}{2}\left(i\theta\sigma \cdot \mathbf{n} - \varsigma\mathbf{K} \cdot \frac{\mathbf{v}}{v}\right)}, \tag{8.60}$$

which is a similarity transformation, the representation (8.58) is equivalent to

$$J^i \to -\frac{1}{2}\sigma^i \ , \ K^i \to -\frac{i}{2}\sigma^i$$

which gives

$$M^i = 0, \ N^i = -\frac{\sigma^i}{2} \tag{8.61}$$

and corresponds to the spinor representation $(0, 1/2)$. The two-component Weyl spinor ξ which transforms as

$$\xi \to \xi' = A\xi = e^{\frac{1}{2}\left(i\theta\sigma \cdot \mathbf{n} + \varsigma\sigma \cdot \frac{\mathbf{v}}{v}\right)}\xi$$

corresponds to the spinor representation $(1/2, 0)$ and the Weyl spinor:

$$\bar{\xi} \equiv i\sigma^2 \xi^*$$

transforms as

$$\bar{\xi} \to \bar{\xi}' = i\sigma^2 A^* \xi^* = \sigma^2 A^* \sigma^2 i\sigma^2 \xi^*$$

$$= \sigma^2 A^* \sigma^2 \bar{\xi} = e^{\frac{1}{2}\left(i\theta\sigma \cdot \mathbf{n} - \varsigma\sigma \cdot \frac{\mathbf{v}}{v}\right)}\bar{\xi}$$

correspond to the spinor representation $(0, 1/2)$.

Accordingly, the spinors

$$\xi = \begin{pmatrix} \xi^1 \\ \xi^2 \end{pmatrix} = \xi^\alpha$$

$$\bar{\xi} = i\sigma^2 \xi^* = \begin{pmatrix} \xi^{*2} \\ -\xi^{*1} \end{pmatrix} = \begin{pmatrix} \xi_1^* \\ \xi_2^* \end{pmatrix} = \xi_\alpha^* = \bar{\xi}_{\dot\alpha} = \begin{pmatrix} \bar\xi_{\dot 1} \\ \bar\xi_{\dot 2} \end{pmatrix}$$

which transforms as

$$\xi^\alpha \to \xi'^\alpha = A^\alpha{}_\beta \xi^\beta$$

$$\bar\xi_{\dot\alpha} \to \bar\xi'_{\dot\alpha} = \left(\sigma^2 A^* \sigma^2\right)_{\dot\alpha}{}^{\dot\beta} \bar\xi_{\dot\beta}$$

are designated representations: $D^{\left(\frac{1}{2},0\right)}$ and $D^{\left(0,\frac{1}{2}\right)}$, respectively. Note that we have used

$$\left(i\sigma^2\right)_{\alpha\beta} \xi^{*\beta} \equiv \varepsilon_{\alpha\beta} \xi^{*\beta} = \xi_\alpha^*$$

$$\varepsilon_{\alpha\beta} = i\sigma^2 = \begin{pmatrix} 0 & 1 \\ -1 & 0 \end{pmatrix}$$

so that

$$\xi_1^* = \xi^{*2},\ \xi_2^* = -\xi^{*1}.$$

As already seen

$$X' = AXA^\dagger \tag{8.62}$$

X represents the four-vector x^μ written as 2×2 matrix. Consider now the quantity $\xi\xi^\dagger$ which transforms as

$$\xi'\xi'^\dagger = A\xi\xi^\dagger A^\dagger$$

$\xi\xi^\dagger$ transforms in the same way as the four vector X, so it can be taken as the basis of the 4-dimensional irreducible representation. It is denoted by

$$D^{(1/2,1/2)} = D^{(1/2,0)} \otimes D^{(0,1/2)}.$$

It gives us spin 1 representation with four components:

$$V^\mu \to V = \begin{pmatrix} V^0 - V^3 & -V^1 + iV^2 \\ -V^1 - iV^2 & V^0 + V^3 \end{pmatrix}.$$

This is reducible, thus

$$D^{(1/2,1/2)}(R) = D^{(1)}(R) \oplus D^{(0)}(R).$$

By the above procedure, one can generate any other representation. First we note that general representations of the Lorentz group are neither parity nor Hermitian conjugate eigenvectors as should be clear from the above discussion. Under parity

$$\left(\frac{1}{2},0\right) \longleftrightarrow \left(0,\frac{1}{2}\right).$$

Thus when parity is considered relevant $\left(\frac{1}{2},0\right) \oplus \left(0,\frac{1}{2}\right)$ gives four component Dirac spinors. There are some more examples like

$$\left(\frac{1}{2},0\right) \otimes \left(\frac{1}{2},0\right) = (0,0) \oplus (1,0)$$

$$\left(0,\frac{1}{2}\right) \otimes \left(0,\frac{1}{2}\right) = (0,0) \oplus (0,1).$$

As is known scalar representation $(0,0)$ is given by antisymmetric product $\frac{1}{2}(\uparrow\downarrow - \downarrow\uparrow)$. Representation $(1,0)$ has 3 independent components and can be represented by an antisymmetric self dual second rank tensor:

$$G_{\mu\nu} = -G_{\nu\mu}$$

$$G_{\mu\nu} = \frac{1}{2}\varepsilon_{\mu\nu\rho\sigma}G^{\rho\sigma}.$$

The representation $(0,1)$ would correspond to that which is anti self-dual:

$$G_{\mu\nu} = -\frac{1}{2}\varepsilon_{\mu\nu\rho\sigma}G^{\rho\sigma}.$$

For example Maxwell tensor $F_{\mu\nu}$ transforms as $(0,1) \oplus (1,0)$ under Lorentz group.

Exercise 1:

For the spinor representation of Lorentz group:

$$S = 1 - \frac{\imath}{2}\epsilon_{\mu\nu}\Sigma^{\mu\nu}$$

$$= 1 - \frac{\imath}{2}\epsilon_{0i}\Sigma^{0i} - \frac{\imath}{2}\epsilon_{i0}\Sigma^{i0} - \frac{\imath}{2}\epsilon_{ij}\Sigma^{ij},$$

$$\epsilon_{0i} = \varsigma_i = -\epsilon_{i0},$$

$$\epsilon_{ij} = \epsilon_{ijn}\omega^n = -\epsilon^{ijn}\omega^n = -\epsilon_{ijn}\omega_n.$$

For spinor representation, $\Sigma^{\mu\nu}$ should involve Pauli matrices. Σ^{ij} is antisymmetric tensor. In terms of Pauli matrices:

$$\Sigma^{ij} = \frac{1}{2}\epsilon^{ijk}\sigma^k$$

and

$$\Sigma^{0i} = -\frac{\imath}{2}\sigma^i,$$

or

$$\Sigma^{0i} = \frac{\imath}{2}\sigma^i.$$

For spinor $\xi\colon (\frac{1}{2}, 0)$, select $\Sigma^{0i} = -\frac{\imath}{2}\sigma^i$. Thus for the spinor $(\frac{1}{2}, 0)$

$$S = 1 - \frac{1}{2}\varsigma_i\sigma^i + \frac{\imath}{4}\epsilon^{\imath jn}\omega^n\epsilon^{\imath jk}\sigma^k$$

$$= 1 + \frac{1}{2}\varsigma\cdot\sigma + \frac{\imath}{2}\omega\cdot\sigma \to e^{\frac{1}{2}(\imath\omega\cdot\sigma + \varsigma\cdot\sigma)} \equiv \Lambda$$

and

$$\sigma^2 S^\star \sigma^2 = 1 - \frac{1}{2}\varsigma\cdot\sigma + \frac{\imath}{2}\omega\cdot\sigma \to e^{\frac{1}{2}(\imath\omega\cdot\sigma - \varsigma\cdot\sigma)}$$

$$= \sigma^2 A^\star \sigma^2$$

$$= \sigma^2 \Lambda^* \sigma^2 \equiv \bar{\Lambda}.$$

Thus

$$\bar{\Lambda}^* = \sigma^2 \Lambda \sigma^2.$$

Thus $A, \sigma^2 A^\star \sigma^2$ and Λ, $\bar{\Lambda}$ are matrix representations of Lorentz group for the spinors $(\frac{1}{2}, 0)$ and $(0, \frac{1}{2})$ respectively.

Exercise 2:

Derive the matrix representations of Lorentz group of spinors, using Dirac γ-matrices and the Dirac spinor Ψ. For Dirac spinor

$$\Sigma^{\mu\nu} = \frac{\imath}{4}(\gamma^\mu\gamma^\nu - \gamma^\nu\gamma^\mu).$$

In the chiral representations of γ-matrices

$$\gamma^\mu = \begin{pmatrix} 0 & \sigma^\mu \\ \bar{\sigma}^\mu & 0 \end{pmatrix},$$

$$\gamma^\mu\gamma^\nu = \begin{pmatrix} \sigma^\mu\bar{\sigma}^\nu & 0 \\ 0 & \bar{\sigma}^\mu\sigma^\nu \end{pmatrix},$$

$$\Sigma^{\mu\nu} = \frac{\imath}{4}\begin{pmatrix} \sigma^\mu\bar{\sigma}^\nu - \sigma^\nu\bar{\sigma}^\mu & 0 \\ 0 & \bar{\sigma}^\mu\sigma^\nu - \bar{\sigma}^\nu\sigma^\mu \end{pmatrix}.$$

Thus

$$\Sigma^{ij} = -\frac{\imath}{4}\begin{pmatrix} \sigma^i\sigma^j - \sigma^j\sigma^i & 0 \\ 0 & \sigma^i\sigma^j - \sigma^j\sigma^i \end{pmatrix}$$

$$= \frac{1}{2}\epsilon^{ijk}\begin{pmatrix} \sigma^k & 0 \\ 0 & \sigma^k \end{pmatrix},$$

$$\Sigma^{0i} = -\Sigma^{i0} = \begin{pmatrix} -\frac{\imath}{2}\sigma^i & 0 \\ 0 & \frac{\imath}{2}\sigma^i \end{pmatrix}.$$

Thus

$$\epsilon_{\mu\nu}\Sigma^{\mu\nu} = 2\epsilon_{0i}\Sigma^{0i} + \epsilon_{ij}\Sigma^{ij}$$

$$= \begin{pmatrix} i\varsigma\cdot\sigma - \omega\cdot\sigma & 0 \\ 0 & -i\varsigma\cdot\sigma - \omega\cdot\sigma \end{pmatrix},$$

and

$$S = 1 - \frac{i}{2}\epsilon_{\mu\nu}\Sigma^{\mu\nu} = \begin{pmatrix} 1 + \frac{i}{2}\omega\cdot\sigma + \frac{1}{2}\varsigma\cdot\sigma & 0 \\ 0 & 1 + \frac{i}{2}\omega\cdot\sigma - \frac{1}{2}\varsigma\cdot\sigma \end{pmatrix}$$

$$\rightarrow \begin{pmatrix} \Lambda_L & 0 \\ 0 & \Lambda_R \end{pmatrix},$$

where

$$\Lambda_L = e^{\frac{1}{2}(i\omega\cdot\sigma + \varsigma\cdot\sigma)} \equiv A,$$

$$\Lambda_R = e^{\frac{1}{2}(i\omega\cdot\sigma - \varsigma\cdot\sigma)} = \sigma^2 A^\star \sigma^2$$

$$= \sigma^2 \Lambda_L^* \sigma^2,$$

$$\Lambda_L = \sigma^2 \Lambda_R^* \sigma^2.$$

Note that

$$\Lambda_L^+ = e^{\frac{1}{2}(-i\omega\cdot\sigma + \varsigma\cdot\sigma)},$$

$$\Lambda_L^{-1} = e^{-\frac{1}{2}(i\omega\cdot\sigma + \varsigma\cdot\sigma)} \neq \Lambda_L^+,$$

$$\Lambda_R^+ \neq \Lambda_R^{-1},$$

$$\Lambda_R^+ = \Lambda_L^{-1}.$$

Thus the spinor representations of Lorentz group are not unitary. Now

$$\Psi_D \equiv \Psi = \begin{pmatrix} \Psi_L \\ \Psi_R \end{pmatrix}$$

$$= \begin{pmatrix} \xi \\ i\sigma^2\chi \end{pmatrix}$$

$$= \begin{pmatrix} \xi \\ \chi^* \end{pmatrix}$$

$$= \begin{pmatrix} \xi^\alpha \\ \bar{\chi}_{\dot\alpha} \end{pmatrix}.$$

$$\Psi_L \rightarrow \Psi_L' = \Lambda_L \Psi_L,$$

$$\xi' = \Lambda_L \xi \equiv A\xi,$$

$$\Psi_R \rightarrow \Psi_R' = \Lambda_R \Psi_R,$$

$$\bar{\chi}' = \Lambda_R \bar{\chi} \equiv (\sigma^2 A^\star \sigma^2)\bar{\chi},$$

$$\xi'^\alpha = (\Lambda_L)^\alpha_\beta \xi^\beta,$$

$$\bar{\chi}_{\dot{\alpha}} = (\Lambda_R)^{\dot{\beta}}_{\dot{\alpha}} \bar{\chi}_{\dot{\beta}}.$$

Unitary operator corresponding to the homogeneous Lorentz transformation:

$$U = e^{-\imath(\omega \cdot \mathbf{J} + \varsigma \cdot \mathbf{K})}.$$

Thus

$$\xi' = U\xi U^\dagger$$

$$= \xi - \imath[\omega \cdot \mathbf{J} + \varsigma \cdot \mathbf{K}, \xi]$$

$$= (1 + \frac{\imath}{2}\omega \cdot \sigma + \frac{1}{2}\varsigma \cdot \sigma)\xi.$$

Hence we have

$$[\mathbf{J}, \xi] = -\frac{1}{2}\sigma\xi,$$

$$[\mathbf{K}, \xi] = \frac{\imath}{2}\sigma\xi.$$

Similarly for $\bar{\chi}$

$$[\mathbf{J}, \bar{\chi}] = -\frac{1}{2}\sigma\bar{\chi},$$

$$[\mathbf{K}, \bar{\chi}] = -\frac{\imath}{2}\sigma\bar{\chi}.$$

8.4   Poincaré Group and Physical States

Define Pauli-Lubanski operator

$$W^\lambda = -\frac{1}{2}\epsilon^{\lambda\alpha\beta\sigma} M_{\alpha\beta} P_\sigma. \tag{8.63}$$

It is a vector and plays the role of the contravariant angular momentum as we shall see later.

Now

$$P_\lambda W^\lambda = 0, \text{ or } \mathbf{P} \cdot \mathbf{W} = P^0 W^0,$$

where

$$W^0 = -\frac{1}{2}\epsilon^{0\alpha\beta\sigma} M_{\alpha\beta} P_\sigma = -\frac{1}{2}\varepsilon^{ijk} M_{ij} P_k = -\frac{1}{2}\varepsilon^{ijk}\varepsilon^{ijl} J^l P_k$$

$$= \mathbf{J} \cdot \mathbf{P} \tag{8.64}$$

and

$$\begin{aligned}
\left[W^\lambda, P^\mu\right] &= -\frac{1}{2}\varepsilon^{\lambda\alpha\beta\sigma}\left[M_{\alpha\beta}P_\sigma, P^\mu\right] \\
&= -\frac{1}{2}\varepsilon^{\lambda\alpha\beta\sigma}\left\{M_{\alpha\beta}\left[P_\sigma, P^\mu\right] + \left[M_{\alpha\beta}, P^\mu\right]P_\sigma\right\} \\
&= -\frac{1}{2}\varepsilon^{\lambda\alpha\beta\sigma}\left\{\eta^{\mu\rho}\left[M_{\alpha\beta}, P_\rho\right]P_\sigma\right\}.
\end{aligned} \tag{8.65}$$

Since [cf. Eq. (8.33)]

$$\left[M_{\alpha\beta}, P_\rho\right]P_\sigma = i\left(P_\alpha\eta_{\beta\rho} - P_\beta\eta_{\alpha\rho}\right)P_\sigma, \tag{8.66}$$

as $\varepsilon^{\lambda\alpha\beta\sigma}$ is antisymmetric in $\alpha\sigma$ or $\beta\sigma$ while $P_\alpha P_\sigma$ and $P_\beta P_\sigma$ are symmetric. Therefore,

$$\left[W^\lambda, P^\mu\right] = 0. \tag{8.67}$$

Now W^λ is a vector under $L_+^\uparrow$, so

$$\left[W^\lambda, M^{\rho\sigma}\right] = -i\left[W^\rho\eta^{\lambda\sigma} - W^\sigma\eta^{\lambda\rho}\right]. \tag{8.68}$$

It is easy to show that [see Problem 8.7]

$$\left[W^\lambda, W^\mu\right] = i\varepsilon^{\lambda\mu\rho\sigma}W_\rho P_\sigma. \tag{8.69}$$

The Lie algebra of Poincaré group has two invariants

$$\begin{aligned}
P^2 &= P_\mu P^\mu, \\
W^2 &= W^\mu W_\mu,
\end{aligned} \tag{8.70}$$

which commute with all the P_μ and $M_{\mu\nu}$. To gain some physical feeling, let us consider rest frame of the particle

$$p^0 = m, \ \mathbf{p} = 0, \tag{8.71}$$

where

$$\begin{aligned}
W^i &= -\frac{1}{2}\varepsilon^{i\alpha\beta\sigma}M_{\alpha\beta}P_\sigma \\
&= -\frac{1}{2}\varepsilon^{i\alpha\beta 0}M_{\alpha\beta}P_0 \\
&= \frac{1}{2}\varepsilon^{ijk}m\varepsilon^{jkl}J^l \\
&= ms^i, \ i = 1,2,3,
\end{aligned} \tag{8.72}$$

and

$$W^0 = 0,$$

which give

$$\frac{W^\mu}{m} = (0, \mathbf{s}) . \tag{8.73}$$

Now little group is defined by a subgroup of Lorentz transformations which leave the momentum of a particle invariant

$$\mathring{\Lambda}^\mu_\nu \mathring{p}^\nu = \mathring{p}^\mu , \tag{8.74}$$

where $\mathring{p}^\mu$ is an eigenvector of operator P^μ.

For an infinitesimal Lorentz transformation

$$\mathring{\Lambda}^\mu_\nu = \eta^{\mu\lambda} \mathring{\Lambda}^\mu_\nu$$

$$= \eta^{\mu\lambda} \left(\eta_{\lambda\nu} + \mathring{\omega}_{\lambda\nu} \right) . \tag{8.75}$$

Then Eq. (8.74) implies

$$\eta^{\mu\lambda} \left(\eta_{\lambda\nu} + \mathring{\omega}_{\lambda\nu} \right) \mathring{p}^\nu = \mathring{p}^\mu ,$$

$$\mathring{\omega}_{\lambda\nu} \mathring{p}^\nu = 0 . \tag{8.76}$$

This latter condition is manifested if we choose

$$\mathring{\omega}_{\lambda\nu} = i\varepsilon^{\lambda\nu\rho\sigma} \theta_\rho \mathring{p}_\sigma = -\mathring{\omega}_{\nu\mu} . \tag{8.77}$$

The little group transformations are therefore described in Hilbert space by the unitary operator

$$U_{\mathring{\Lambda}} = \exp\left(-\frac{1}{2} \mathring{\omega}^{\mu\nu} M_{\mu\nu} \right)$$

$$= \exp\left(i\theta_\rho W^\rho \right) . \tag{8.78}$$

Thus W^ρ plays the role of contravariant angular momentum.

The Casimir invariants $P^2 = P_\mu P^\mu$ and $W^2 = W^\mu W_\mu$ are also Lorentz invariants, and as such they can be evaluated in any frame. In particular in the rest frame of particle

$$\mathring{p}^\mu \to (m, \mathbf{0}) , \quad W^\mu \to (0, \mathbf{s}) ,$$

and so little group of Poincaré group associated with a massive particle is the rotation group $O(3)$ with eigenvalues of the Casimir invariants

$$P^2 = m^2$$

and

$$W^2 = -ms(s+1) . \tag{8.79}$$

In accordance with Schur's lemma, irreducible representations of Poincaré group are characterized by eigenvalues of the two Casimir invariants, which as seen above correspond to mass m and spin s. From the generators and the quantities W^λ, we choose a maximal abelian subset P^λ and W^3, the eigenvalues of these operators serve to distinguish the different sates corresponding to a given irreducible representation. Such states may therefore denoted by $|p, w^3; m, s\rangle$, where

$$P^2|p, w^3; m, s\rangle = m^2|p, w^3; m, s\rangle,$$

$$W^2|p, w^3; m, s\rangle = -m^2 s\,(s+1)\,|p, w^3; m, s\rangle,$$

$$P^\mu|p, w^3; m, s\rangle = p^\mu|p, w^3; m, s\rangle,$$

$$W^3|p, w^3; m, s\rangle = w^3|p, w^3; m, s\rangle. \tag{8.80}$$

The physical interpretation of these states is that p^μ represents the 4-momentum of the state, while w^3 is m times the third component of its spin in the Lorentz frame in which the 3-momentum vanishes.

Thus we can say that states of a physical system are described by vectors in a linear vector space which is the representation space for the Poincaré group.

For a massless particle, we can write from $W^\mu P_\mu = 0$ and $P^2 = P^\mu P_\mu = 0$

$$W^\mu = \lambda P^\mu. \tag{8.81}$$

Then from Eq. (8.64)

$$\lambda = \frac{W^0}{P^0} = \frac{P^0}{|\mathbf{P}|^2}\mathbf{J}\cdot\mathbf{P} = \mathbf{J}\cdot\mathbf{n},$$

$$\mathbf{P} = \mathbf{n}P^0. \tag{8.82}$$

Thus one can see that for a massless particle λ is the helicity with eigenvalues $= \pm s$

$$s = \begin{cases} +\frac{1}{2} & \text{for spinor} \\ 1 & \text{for vector (photon)}. \end{cases} \tag{8.83}$$

Lastly can one combine an internal symmetry generated by T^A

$$[T^A, T^B] = if^{AB}_C T^C,$$

with Poincaré group in a non-trivial way?

The answer is no; there is a no go theorem of O' Raifeartaigh [Phys. Rev. Lett. **14** (1965) 575; Phys. Rev. **139** (1965) B1052] and Colemann and Mandula [Phys. Rev. **159** (1967) 1251]: The combined symmetry is always a direct product: Poincaré group $\otimes$ G_{internal}, i.e.,

$$[T^A,\ \text{Poincaré}] = 0.$$

A simple example can be given to exhibit the difficulty by considering $SU(2)$ isospin group [where proton p and neutron n form an isospin doublet with $m_p = m_n$]. Let us assume that

$$\left[\tau^+, P^2\right]|n\rangle \neq 0,$$

$\frac{\tau_i}{2}$ is isospin generator $\tau^+|n\rangle = |p\rangle$.

Now

$$\tau^+ P^2 |n\rangle - P^2 \tau^+ |n\rangle$$

$$= \tau^+ m_n^2 |n\rangle - P^2 |p\rangle = \left(m_n^2 - m_p^2\right)|p\rangle \implies m_n^2 \neq m_p^2,$$

which contradicts the basic assumption.

8.5   Scale Invariance

In order to discuss scale invariance, let us first consider the dimensions of an operator. Now the action

$$S = \int \mathcal{L} \, d^4 x$$

is dimensionless. Therefore $\mathcal{L}$ must have dimensions $\frac{1}{l^4} = (mass)^4$ or simply dimension 4.

We now consider the kinetic energy term of free Lagrangian for neutral scalar field

$$\mathcal{L} = \partial_\mu \phi \partial^\mu \phi,$$

where $\partial_\mu = \frac{\partial}{\partial x^\mu}$ has dimension of $\frac{1}{l} = mass$ and hence ϕ has dimension $l = -1$ or dimension 1.

The Lagrangian for a spin $1/2$ field is

$$\mathcal{L} = \bar{\psi}(x) \left[i\gamma^\mu \partial_\mu\right] \psi(x).$$

The field ψ has $l = -\frac{3}{2}$ or dimension $\frac{3}{2}$.

For vector field Lagrangian, it is

$$\mathcal{L} = -\frac{1}{4} F^{\mu\nu} F_{\mu\nu},$$

$F_{\mu\nu}$ has dimension 2 and A_μ has dimension 1.

The scale dimension l should not be confused with "dimensions" of ordinary dimensional analysis. The scale dimensionality is defined by the behavior of these fields under scale transformation:

$$x^\mu \to x'^\mu = \rho x^\mu,$$

$$\phi(x) \to \phi'(x') = \rho^l \phi(x) = \rho^l \phi\left(\rho^{-1} x'\right), \tag{8.84}$$

where l is dimension characteristic of ϕ, and is called the scale dimension of ϕ.

In quantum field theory, equal time commutation relation should maintain the invariance under the above transformation, e.g. for a scalar field

$$\left[\dot{\phi}(\mathbf{x}, t), \phi(\mathbf{y}, t)\right] \rightarrow \left[\rho^{l-1}\dot{\phi}(\mathbf{x}, t), \rho^{l}\phi(\mathbf{y}, t)\right]$$

$$= \rho^{2l-1}\left[\dot{\phi}(\mathbf{x}, t), \phi(\mathbf{y}, t)\right]$$

$$\rightarrow -\frac{i}{\rho^3}\delta(\mathbf{x} - \mathbf{y}). \tag{8.85}$$

Invariance requires

$$\rho^{2l-1} = \frac{1}{\rho^3}, \, l = -1. \tag{8.86}$$

For spin $\frac{1}{2}$

$$\left[\psi(\mathbf{x}, t), \bar{\psi}(\mathbf{y}, t)\right]_+ = \frac{1}{\rho^3}\delta(\mathbf{x} - \mathbf{y}),$$

$$\rho^{2l} = \frac{1}{\rho^3}, \, l = -3/2. \tag{8.87}$$

The values of l for the boson and fermion fields coincide with the "canonical" dimensions of these fields within the framework of the usual dimensional analysis. Consider now boson mass term $m^2\phi^2$ which has ordinary dimension $l = -4$ but the scale dimension $l = -2$. Similarly fermion mass term $m\bar{\psi}\psi$ has ordinary dimension $l = -4$ but the scale dimension $l = -3$.

8.5.1   *Scale Transformation*

$$x'^{\mu} = \rho x^{\mu} \tag{8.88}$$

or

$$x'^{\mu} = x^{\mu}(1 + \lambda) \text{ for infinitesimal case.}$$

The corresponding unitary transformation is

$$U(\rho)\phi(x)U^{\dagger}(\rho) = \phi'(x) = \rho^{l}\phi\left(\rho^{-1}x\right), \tag{8.89}$$

where we have used Eq. (8.84).

For an infinitesimal transformation

$$U = 1 + i\lambda D, \tag{8.90}$$

the above equation implies

$$(1 + i\lambda D) \phi(x) (1 - i\lambda D) = (1 + l\lambda) \phi(x^\mu - \lambda x^\mu)$$

$$= (1 + l\lambda) [\phi(x) - \lambda x^\mu \partial_\mu \phi(x)] . \quad (8.91)$$

Hence we have

$$[D, \phi(x)] = [-il + ix^\mu \partial_\mu] \phi(x) \quad (8.92)$$

and D is called the dilation operator: $D \sim ix^\mu \partial_\mu$.

8.5.2 *Conformal Group*

Consider a transformation

$$x^\mu \to x'^\mu = x^\mu + \epsilon^\mu(x),$$

such that

$$ds^2 = \eta_{\mu\nu} dx^\mu dx^\nu$$

$$\to \eta_{\mu\nu} dx'^\mu dx'^\nu$$

$$= F(x)\eta_{\mu\nu} dx^\mu dx^\nu . \quad (8.93)$$

Now

$$dx'^\mu = dx^\mu + \partial_\rho \epsilon^\mu(x) dx^\rho,$$

$$dx'^\nu = dx^\nu + \partial_\sigma \epsilon^\nu(x) dx^\sigma,$$

$$\eta_{\mu\nu} dx'^\mu dx'^\nu = \begin{bmatrix} \eta_{\mu\nu} dx^\mu dx^\nu + \eta_{\mu\nu} \left(\partial_\sigma \epsilon^\nu(x) dx^\sigma \right) dx^\mu \\ + \eta_{\mu\nu} \left(\partial_\rho \epsilon^\mu(x) dx^\rho \right) dx^\nu \end{bmatrix}$$

$$= [\eta_{\mu\nu} dx^\mu dx^\nu + (\partial_\mu \epsilon_\nu(x) + \partial_\nu \epsilon_\mu(x)) dx^\mu dx^\nu] . \quad (8.94)$$

Equation (8.93) requires that

$$\eta_{\mu\nu} + \partial_\mu \epsilon_\nu(x) + \partial_\nu \epsilon_\mu(x) = F(x)\eta_{\mu\nu} . \quad (8.95)$$

Multiply Eq. (8.95) with $\eta^{\mu\nu}$, i.e., taking a trace ($\eta^{\mu\nu}\eta_{\mu\nu} = d$), we get

$$2\partial \cdot \epsilon(x) = d[-1 + F(x)],$$

or

$$F(x) = \frac{2}{d}\partial \cdot \epsilon(x) + 1. \quad (8.96)$$

Thus

$$\partial_\mu \epsilon_\nu(x) + \partial_\nu \epsilon_\mu(x) = \frac{2}{d}\partial \cdot \epsilon(x)\eta_{\mu\nu},$$

where d is the dimension of space time. Operation by ∂^μ on both sides gives

$$\Box\epsilon_\nu(x) + \left(1 - \frac{2}{d}\right)\partial_\nu\left(\partial \cdot \epsilon(x)\right) = 0. \tag{8.97}$$

The quantity $\left(1 - \frac{2}{d}\right)$ does not vanish for $d > 2$. We may parametrize $\epsilon^\mu(x)$ as follows

$$\epsilon^\mu(x) = a^\mu + \omega^{\mu\nu}x_\nu + \lambda x^\mu + c^\mu x^2 + \left(b^\kappa x_\kappa\right)x^\mu. \tag{8.98}$$

Then

$$\partial \cdot \epsilon = \partial_\mu\left[c^\mu x^2 + b^\kappa x_\kappa x^\mu\right] + \lambda\delta^\mu_\mu$$
$$= 2c \cdot x + (d+1)b \cdot x + \lambda d,$$
$$\partial_\nu\partial \cdot \epsilon = 2c^\mu\eta_{\mu\nu} + (d+1)b^\kappa\eta_{\nu\kappa}$$
$$= 2c_\nu + (d+1)\,b_\nu,$$
$$\Box\epsilon^\mu = \partial_\lambda\partial^\lambda \cdot \epsilon^\mu = \partial^\lambda\left\{\begin{matrix}\omega^{\mu\nu}\eta_{\nu\lambda} + \lambda\delta^\mu_\lambda + c^\mu 2x_\lambda \\ +b^\kappa\left(\eta_{\lambda\kappa}x^\mu + x_\kappa\delta^\mu_\lambda\right)\end{matrix}\right\}$$
$$= 2dc^\mu + 2b^\mu. \tag{8.99}$$

Substituting in Eq. (8.97), we get

$$(2dc_\nu + 2b_\nu) + \left(\left(1 - \frac{2}{d}\right)(2c_\nu + (d+1)\,b_\nu)\right) = 0, \tag{8.100}$$

which we can rewrite as

$$(2d + 2 - 4/d)\,c_\nu + (d + 1 - 2/d)\,b_\nu = 0, \tag{8.101}$$

$$b_\nu = -2c_\nu.$$

Thus

$$\epsilon^\mu(x) = a^\mu + \omega^{\mu\nu}x_\nu + \lambda x^\mu + c^\mu x^2 - 2c \cdot x x^\mu. \tag{8.102}$$

We can now identify various terms in $\epsilon^\mu(x)$:

$$a^\mu \;:\; \text{translation}$$
$$\omega^{\mu\nu} = -\omega^{\nu\mu} \qquad L^\uparrow_+ \tag{8.103}$$
$$\lambda \;:\; \text{scale transformation}$$
$$c^\mu x^2 - 2c \cdot x x^\mu \;:\; \text{special conformal transformation.}$$

Let us now specialize to four dimensions, where we have

$$(4 + 6 + 1 + 4) = 15 \text{ parameters.}$$

The corresponding 15 generators are

$$P^\mu \sim -i\partial^\mu,$$
$$J^{\mu\nu} \sim L^{\mu\nu} = i\left(x^\mu\partial^\nu - x^\nu\partial^\mu\right),$$
$$D \sim ix^\mu\partial_\mu, \tag{8.104}$$
$$\bar{K}^\mu.$$

Consider now the special conformal transformation

$$x'^\mu \to x^\mu \to x^\mu + c^\mu x^2 - (2c \cdot x)\, x^\mu, \tag{8.105}$$

the corresponding unitary transformation

$$U = 1 + ic^\mu \bar{K}_\mu, \tag{8.106}$$

and

$$\phi(x) \to \phi'(x') = U\phi(x')U^\dagger$$
$$U^\dagger \phi(x)U = \phi(x') = \phi\left(x^\mu + c^\mu x^2 - (2c \cdot x)\, x^\mu\right)$$
$$\phi(x) - ic^\mu\left[\bar{K}_\mu, \phi(x)\right] = \phi(x) + \left(c^\mu x^2 - 2c \cdot xx^\mu\right)\frac{\partial\phi}{\partial x^\mu}.$$

Thus

$$\bar{K}_\mu \sim i\left(x^2\partial_\mu - 2x_\mu x^\nu\partial_\nu\right). \tag{8.107}$$

The 15 generators $P_\mu, J_{\mu\nu}, D, \bar{K}_\mu$ generate conformal group. The commutation relations can be easily calculated

$$[P_\mu, D] = iP_\mu,$$
$$[D, M_{\mu\nu}] = 0,$$
$$[D, \bar{K}_\mu] = i\bar{K}_\mu,$$
$$[\bar{K}_\mu, \bar{K}_\nu] = 0,$$
$$[\bar{K}_\mu, P_\nu] = 2i\eta_{\mu\nu}D + 2iM_{\mu\nu},$$
$$[\bar{K}_\rho, M_{\mu\nu}] = i\left(\eta_{\rho\mu}\bar{K}_\nu - \eta_{\rho\nu}\bar{K}_\mu\right). \tag{8.108}$$

8.6 Energy Momentum Tensor $T^{\mu\nu}$

It is

(i) symmetric $T_{\mu\nu} = T_{\nu\mu}$

(ii) conserved $\partial^\mu T_{\mu\nu} = 0$

(iii) matrix element between physical states may be finite.

Compare

$$[D, \phi(x)] = (ix^\mu \partial_\mu - il)\phi(x),$$

with

$$[P_\mu, \phi(x)] = -i\partial_\mu \phi(x),$$

and in analogy with

$$P_\mu = \int d^3x\, T_{\mu 0},$$

$$D = \int d^3x\, x^\mu T_{\mu 0}$$

$$= \int d^3x\, x^i T_{i0} + \int d^3x\, x^0 T_{00}$$

$$= \int d^3x\, x^i T_{i0} + P_0 t,$$

so that

$$\frac{dD}{dt} = P_0 + \int d^3x\, x^i \frac{dT_{i0}}{dt}. \tag{8.109}$$

Now by using the conservation of $T_{\mu\nu}$, i.e., $\partial^\mu T_{\mu\nu} = 0$, we have for $\nu = i$

$$\frac{\partial T_{0i}}{\partial t} = -\partial^j T_{ji}. \tag{8.110}$$

Substituting in Eq. (8.109), we obtain

$$\frac{dD}{dt} = P_0 + \int d^3x\, x^i \left(-\partial^j T_{ji}\right)$$

$$= P_0 - \int d^3x\, \partial^i \left(x^j T_{ij}\right) + \int d^3x\, \eta^{ij} T_{ij}. \tag{8.111}$$

Due to Gauss's theorem $\int d^3x\, \partial^i \left(x^j T_{ij}\right) = 0$, so that

$$\frac{dD}{dt} = \int d^3x\, T^\mu_\mu. \tag{8.112}$$

Thus it is possible to define a dilation current

$$S_\mu = x^\nu T_{\mu\nu},$$

$$\partial^\mu S_\mu = \eta^{\mu\nu} T_{\mu\nu} + x^\nu \partial^\mu T_{\mu\nu}. \tag{8.113}$$

In the limit of scale invariance

$$T^\mu_\mu = 0, \partial^\mu S_\mu = 0, \tag{8.114}$$

according to our choice of D in order that $\frac{dD}{dt} = 0$.

Let us consider the matrix elements of $T_{\mu\nu}$ between one particle states:

$$\langle p'|T_{\mu\nu}|p\rangle = \eta_{\mu\nu}F_1\left(q^2\right)+P_\mu P_\nu F_2\left(q^2\right)+q_\mu q_\nu F_3\left(q^2\right)+(q_\mu P_\nu + q_\nu P_\mu)\,F_4\left(q^2\right),$$

(8.115)

where $q = \frac{p-p'}{2}$ and $P = \frac{p+p'}{2}$ and it is understood that we are taking the spin average in the matrix elements. If we contract the above equation with $q^\mu, \partial^\mu T_{\mu\nu} = 0$ gives

$$q^\mu\langle p'|T_{\mu\nu}|p\rangle = q_\nu F_1\left(q^2\right) + \frac{p^2 - p'^2}{4}F_2\left(q^2\right) + q^2 q_\nu F_3 + q^2 p_\nu F_4,\quad (8.116)$$

so that with $p^2 = p'^2 = m^2$ and $q^2 = 0$, $F_1\left(0\right) = 0$. Thus for particles on mass shell,

$$\langle p|T_{\mu\nu}|p\rangle = p_\mu p_\nu F_2\left(0\right).$$

(8.117)

Taking the trace by multiplying the above equation with $g^{\mu\nu}$, we obtain

$$\langle p|T^\mu_\mu|p\rangle = m^2 F_2\left(0\right).$$

(8.118)

Thus $T^\mu_\mu = 0$ implies $m^2 = 0$, i.e., for example proton mass is zero. In the limit $m_{quark} \to 0$, Quantum Chromodynamics (QCD) is scale invariance implying $T^\mu_\mu = 0$, i.e., the proton is massless. Quantum radiative corrections destroy the classical symmetry in order that proton to have mass.

8.7 Supersymmetry (SUSY)

The simplest example of supersymmerty is provided by simple harmonic oscillator for which the Hamiltonian $[\hbar = 1, m = 1, \omega = 1]$

$$\begin{aligned}
H_B &= \frac{1}{2}\left(p^2 + q^2\right)\\
&= \frac{1}{\sqrt{2}}\left(q - ip\right)\frac{1}{\sqrt{2}}\left(q + ip\right) - \frac{i}{2}\left[q,p\right]\\
&= \hat{a}^\dagger\hat{a} + \frac{1}{2},
\end{aligned}$$

(8.119)

where $\hat{a}^\dagger$ and $\hat{a}$ are creation and annihilation operators, and satisfy

$$\left[\hat{a}, \hat{a}^\dagger\right] = 1.$$

(8.120)

The eigenstates of H_B are $|0\rangle_B, |1\rangle_B, |2\rangle_B, \ldots$.

$$\hat{a}|0\rangle_B = 0, \text{ for ground state}$$

$$_B\langle 0|H_B|0\rangle_B = \frac{1}{2}, \text{ ground state energy.}$$

(8.121)

Normalized eigenstates are

$$|n\rangle_B = \frac{1}{\sqrt{n!}} \left(\hat{a}^\dagger\right)^n |0\rangle_B,$$

$$H_B |n\rangle_B = \left(n + \frac{1}{2}\right) |n\rangle_B. \tag{8.122}$$

We now introduce a fermion oscillator which consists of only two states corresponding to spin $\frac{1}{2}$:

$$|\uparrow\rangle = \begin{pmatrix} 1 \\ 0 \end{pmatrix},$$

$$|\downarrow\rangle = \begin{pmatrix} 0 \\ 1 \end{pmatrix}. \tag{8.123}$$

Creation and annihilation operators are

$$d = \sigma_- = \frac{\sigma_1 - i\sigma_2}{2}$$

$$= \begin{pmatrix} 0 & 0 \\ 1 & 0 \end{pmatrix},$$

$$d^\dagger = \sigma_+ = \begin{pmatrix} 0 & 1 \\ 0 & 0 \end{pmatrix}, \tag{8.124}$$

$$d^\dagger |\downarrow\rangle = |\uparrow\rangle,$$

$$d|\downarrow\rangle = 0, \tag{8.125}$$

$$\{d, d^\dagger\} \equiv dd^\dagger + d^\dagger d = 1. \tag{8.126}$$

The Hamiltonian for fermion oscillator is

$$H_F = \frac{1}{2}\sigma_3 = d^\dagger d - \frac{1}{2}, \tag{8.127}$$

and eigenstates of H_F are: $|0\rangle_F$ and $d^\dagger |0\rangle_F$ so that $d|0\rangle_F = 0$.

SUSY oscillator is the combination of boson and fermion oscillators

$$H = H_B + H_F$$

$$= a^\dagger a + d^\dagger d, \tag{8.128}$$

and eigenstates are

$$|0\rangle, \quad |1\rangle = a^\dagger |0\rangle, \quad |2\rangle = \left(a^\dagger\right)^2 |0\rangle, \ldots$$

$$|1\rangle = d^\dagger |0\rangle, \quad |2\rangle = d^\dagger |1\rangle, \ldots$$

Note the degeneracy of states in supersymmetry harmonic oscillator. The degeneracy in energy indicates that there is a symmetry.

Define operators

$$Q = \sqrt{2}a^\dagger d, \ Q^\dagger = \sqrt{2}ad^\dagger,$$

Q is no longer a pure bosonic or fermionic object and Q and $Q^\dagger$ provide the simplest supersymmetric algebra, which one can easily check, is

$$\{Q, Q\} = \{Q^\dagger, Q^\dagger\} = 0,$$
$$\{Q, Q^\dagger\} = 2H, \tag{8.129}$$
$$[Q, H] = 0, \ [Q^\dagger, H] = 0.$$

These relations explain the degeneracy mentioned above. The above consideration show that any enlargement of space-time graph require generators which anticommute.

8.8   SUSY Quantum Mechanics

We first consider Lagrangian formalism for classical system described by generalized coordinates

$$q(t), \dot{q}(t), \ \psi(t), \bar{\psi}(t).$$

Here q''s are bosonic coordinates and ψ is anticommuting c-number called Grassmann numbers satisfying the algebra

$$\psi_1 \psi_2 = -\psi_2 \psi_1, \ (\psi_i)^2 = 0. \tag{8.130}$$

Consider the Lagrangian

$$\mathcal{L} = \frac{1}{2}\dot{q}^2 - \frac{1}{2}[V(q)]^2 + \frac{i}{2}\left(\bar{\psi}\dot{\psi} - \dot{\bar{\psi}}\psi\right) - V'(q)\bar{\psi}\psi, \tag{8.131}$$

where

$$V' = \frac{\partial V}{\partial q}.$$

The Hamiltonian of the system is

$$\mathcal{H} = \sum_j p_j \dot{q}_j - \mathcal{L}. \tag{8.132}$$

Now

$$\frac{\partial \mathcal{L}}{\partial \dot{q}} = \dot{q}, \ \frac{\partial \mathcal{L}}{\partial \dot{\psi}} = \frac{i}{2}\bar{\psi}, \ \frac{\partial \mathcal{L}}{\partial \dot{\bar{\psi}}} = -\frac{i}{2}\psi, \tag{8.133}$$

so that

$$\mathcal{H} = \frac{1}{2}\dot{q}^2 + \frac{1}{2}[V(q)]^2 + V'(q)\bar{\psi}\psi. \tag{8.134}$$

For a Harmonic oscillator $V(q) = q$, $V'(q) = 1$. The Lagrangian equation

$$\frac{d}{dt}\left(\frac{\partial \mathcal{L}}{\partial \dot{q}}\right) - \frac{\partial \mathcal{L}}{\partial q} = 0, \tag{8.135}$$

i.e.,

$$\ddot{q} + V(q)V'(q) + V''(q)\bar{\psi}\psi = 0. \tag{8.136}$$

Similarly,

$$i\dot{\bar{\psi}} + V'(q)\bar{\psi} = 0, \quad -i\dot{\psi} + V'(q)\psi = 0. \tag{8.137}$$

Using this equation, one can show that [see Problem]

$$\delta\mathcal{L} = \frac{d}{dt}\left[\dot{q}\delta q + \frac{i}{2}(\bar{\psi}\delta\psi - \delta\bar{\psi}\psi)\right]. \tag{8.138}$$

Under transformations (which close properly):

$$\delta q = \epsilon\psi + \bar{\psi}\bar{\epsilon},$$

$$\delta\bar{\psi} = -(i\dot{q} + V(q))\,\epsilon,$$

$$\delta\psi = (i\dot{q} - V(q))\,\bar{\epsilon},$$

where ϵ is anticommuting (constant) parameter (cf. Eq. (8.138))

$$\delta\mathcal{L} = \frac{i}{2}\frac{d}{dt}\left[\{-\epsilon(i\dot{q} - V(q))\psi - \bar{\psi}(i\dot{q} + V(q))\bar{\epsilon}\}\right]. \tag{8.139}$$

Thus the invariance of the Lagrangian under the above transformation leads to conserved charges ($p = \dot{q}$ in classical mechanics)

$$Q = (-ip + V(q))\psi, \quad Q^\dagger = \bar{\psi}(ip + V(q)). \tag{8.140}$$

Using

$$\psi\bar{\psi} = \frac{1}{2}\{\psi, \bar{\psi}\} + \frac{1}{2}[\psi, \bar{\psi}]$$

$$\bar{\psi}\psi = \frac{1}{2}\{\psi, \bar{\psi}\} - \frac{1}{2}[\psi, \bar{\psi}]$$

it can be easily shown that

$$\{Q, Q^\dagger\} = (p^2 + V^2)\{\psi, \bar{\psi}\} - V'(q)[\psi, \bar{\psi}]. \tag{8.141}$$

In our case

$$\psi = \frac{1}{2}\sigma^-, \quad \bar{\psi} = \frac{1}{2}\sigma^+.$$

Thus

$$\{Q, Q^\dagger\} = \begin{pmatrix} p^2 + V^2 + V' & 0 \\ 0 & p^2 + V^2 - V' \end{pmatrix} = 2H. \tag{8.142}$$

8.8.1 *SUSY Vacuum*

It is defined by

$$Q\Psi = Q^\dagger \bar\Psi = 0,$$

$$\Psi = c_1 \begin{pmatrix} 0 \\ e^{-W(q)} \end{pmatrix} + c_2 \begin{pmatrix} e^{W(q)} \\ 0 \end{pmatrix}. \tag{8.143}$$

$W(q)$ is called super-potential, $W'(q) = V(q)$. Now in our case Q and $Q^\dagger$ are given in Eq. (8.140). Using these values of Q and $Q^\dagger$ given in Eq. (8.140), it is easy to verify that

$$Q\Psi = Q^\dagger \bar\Psi = 0. \tag{8.144}$$

8.9 Super Lie Algebra

We have already seen that the Poincaré group generators are

$$P_\mu = -i\partial_\mu,$$

$$J_{\mu\nu} = L_{\mu\nu} + \frac{1}{2}\Sigma_{\mu\nu}$$

$$= i\left(x_\mu \partial_\nu - x_\nu \partial_\mu\right) + \frac{1}{2}\Sigma_{\mu\nu}, \tag{8.145}$$

where $\Sigma_{\mu\nu}$ is spin matrix. We now introduce Clifford algebra in 4 dimensions

$$\{\gamma^\mu, \gamma^\nu\} = 2\eta^{\mu\nu},$$

$$\eta^{\mu\nu} = \eta_{\mu\nu} = (1, -1, -1, -1),$$

$$\gamma^\mu = \begin{pmatrix} 0 & \sigma^\mu \\ \bar\sigma^\mu & 0 \end{pmatrix},$$

$$\gamma^0 = \begin{pmatrix} 0 & 1 \\ 1 & 0 \end{pmatrix},$$

$$\gamma^5 = \begin{pmatrix} -1 & 0 \\ 0 & 1 \end{pmatrix}, \tag{8.146}$$

$$\{\gamma^\mu, \gamma^5\} = 0,$$

$$\Sigma^{\mu\nu} = \frac{i}{2}[\gamma^\mu, \gamma^\nu]. \tag{8.147}$$

The Charge conjugation matrix C is defined as

$$C = i\gamma^0 \gamma^2$$

$$= \begin{pmatrix} -\epsilon & 0 \\ 0 & \epsilon \end{pmatrix}, \tag{8.148}$$

where

$$\epsilon = i\sigma^2 = \begin{pmatrix} 0 & 1 \\ -1 & 0 \end{pmatrix}, \tag{8.149}$$

$$C = -C^T = -C^{-1} = -C^\dagger,$$

$$C\gamma^\mu C^{-1} = -(\gamma^\mu)^T. \tag{8.150}$$

Dirac spinor is denoted by ψ and Dirac conjugate is $\bar{\psi} = \psi^\dagger \gamma^0$,

$$\psi^c = C\bar{\psi}^T = C\gamma^0\psi^*. \tag{8.151}$$

The Majorana spinor which is self conjugate is

$$\psi_M = \psi_M^C = C\bar{\psi}_M^T,$$

and thus

$$\bar{\psi}_M = \psi_M^\dagger \gamma^0 = \psi_M^T C. \tag{8.152}$$

Bilinear expressions for Majorana spinors are

$$\bar{\psi}_M (...) \psi_M = \psi_M^T C (...) \psi_M, \tag{8.153}$$

where $(...)$ denote a combination of γ matrices. Now

$$(C\gamma^\mu)^T = C\gamma^\mu, \tag{8.154}$$

which implies

$$(C\gamma^\mu)_{\beta\alpha} = (C\gamma^\mu)_{\alpha\beta}, \tag{8.155}$$

so that for the current

$$\begin{aligned} J^\mu &= \bar{\psi}_M \gamma^\mu \psi_M \\ &= \psi_M^T C\gamma^\mu \psi_M \\ &= \psi_{M\alpha}^T (C\gamma^\mu)_{\alpha\beta} \psi_{M\beta}. \end{aligned} \tag{8.156}$$

But $(C\gamma^\mu)_{\alpha\beta}$ is symmetric while ψ''s are antisymmetric and so $J^\mu = 0$. This is an expression of the fact that the Majorana particles have no charge and so electromagnetic current $J^\mu = 0$.

8.9.1 *Two-component Spinors*

We have already introduced the two-component spinors in sec. 7.3

$$\xi^\alpha \to \xi'^\alpha = A^\alpha{}_\beta \xi^\beta \qquad \xi = \begin{pmatrix} \xi^1 \\ \xi^2 \end{pmatrix},$$

$$\bar{\xi}^{\dot\alpha} \to \bar{\xi}'^{\dot\alpha} = A^{\dot\alpha}{}_{\dot\beta} \bar{\xi}^{\dot\beta} \qquad \xi^* = \begin{pmatrix} \xi^{\dot 1} \\ \xi^{\dot 2} \end{pmatrix} \equiv \bar{\xi}^{\dot\alpha}. \tag{8.157}$$

We now write 4-component Dirac spinor as

$$\Psi = \begin{pmatrix} \Psi_L \\ \Psi_R \end{pmatrix}, \tag{8.158}$$

where in chiral representation

$$\gamma^5 = \begin{pmatrix} -1 & 0 \\ 0 & 1 \end{pmatrix}, \ \gamma^0 = \begin{pmatrix} 0 & 1 \\ 1 & 0 \end{pmatrix}, \tag{8.159}$$

and

$$\Psi_L = \frac{1 - \gamma^5}{2} \Psi = \begin{pmatrix} \xi \\ 0 \end{pmatrix},$$

$$\Psi_R = \frac{1 + \gamma^5}{2} \Psi = \begin{pmatrix} 0 \\ \eta \end{pmatrix} \tag{8.160}$$

are left-handed and right-handed spinors. The spinors ξ and η have two components each and are called Weyl spinors. Thus we can write 4-component Dirac spinor as $\begin{pmatrix} \xi \\ \eta \end{pmatrix}$. Now

$$\Psi^c = C\bar{\Psi}^T = C\left(\Psi^\dagger \gamma^0\right)^T$$
$$= C\gamma^{0T}\Psi^* = i\gamma^0\gamma^2\gamma^0\Psi^* = -i\gamma^2\Psi^*. \tag{8.161}$$

Thus

$$\Psi^c = \begin{pmatrix} \xi^c \\ \eta^c \end{pmatrix} = \begin{pmatrix} 0 & -i\sigma^2 \\ i\sigma^2 & 0 \end{pmatrix} \begin{pmatrix} \xi^* \\ \eta^* \end{pmatrix}. \tag{8.162}$$

Therefore,

$$\xi^c = -i\sigma^2 \eta^*$$
$$\eta^c = i\sigma^2 \xi^*. \tag{8.163}$$

Hence

$$\eta = i\sigma^2 \xi^{c*}. \tag{8.164}$$

It is convenient to introduce the following lowering and raising metric

$$\epsilon_{\alpha\beta} = \epsilon = i\sigma^2 = \begin{pmatrix} 0 & 1 \\ -1 & 0 \end{pmatrix} = \epsilon_{\dot\alpha\dot\beta},$$

$$\epsilon^{\alpha\beta} = \begin{pmatrix} 0 & -1 \\ 1 & 0 \end{pmatrix} = \epsilon^{\dot\alpha\dot\beta}. \tag{8.165}$$

ϵ transforms a contravariant vector into a covariant vector. Thus

$$\xi_\alpha = \epsilon_{\alpha\beta}\xi^\beta, \quad \xi_1 = \xi^2,$$

$$\xi_{\dot\alpha} = \epsilon_{\dot\alpha\dot\beta}\xi^{\dot\beta}, \quad \xi_{\dot2} = -\xi^{\dot1}, \tag{8.166}$$

$$\psi = \begin{pmatrix} \xi \\ \epsilon\xi^{*c} \end{pmatrix} = \begin{pmatrix} \xi \\ i\sigma^2\xi^{*c} \end{pmatrix}. \tag{8.167}$$

Sometimes it is convenient to write right-handed spinor in terms of left-handed antiparticle spinor $\chi = \xi^c$. Thus

$$\psi_D = \begin{pmatrix} \xi \\ i\sigma^2\chi^* \end{pmatrix}, \quad \psi_M = \begin{pmatrix} \xi \\ i\sigma^2\xi^* \end{pmatrix}. \tag{8.168}$$

Now

$$i\sigma^2\chi^* = \begin{pmatrix} 0 & 1 \\ -1 & 0 \end{pmatrix} \begin{pmatrix} \chi^{1*} \\ \chi^{2*} \end{pmatrix}$$

$$= \begin{pmatrix} \chi^{2*} \\ -\chi^{1*} \end{pmatrix}$$

$$= \begin{pmatrix} \chi_1^* \\ \chi_2^* \end{pmatrix} = \chi_\alpha^* \equiv \bar\chi_{\dot\alpha} \tag{8.169}$$

where we have used Eq. (8.166). Thus

$$\psi_D = \begin{pmatrix} \xi^\alpha \\ \bar\chi_{\dot\alpha} \end{pmatrix}, \tag{8.170}$$

$$\psi_M = \begin{pmatrix} \xi^\alpha \\ \bar\xi_{\dot\alpha} \end{pmatrix}, \tag{8.171}$$

$$\bar\psi_D = \begin{pmatrix} \chi_\alpha & \bar\xi^{\dot\alpha} \end{pmatrix}, \tag{8.172}$$

$$\bar\psi_M = \begin{pmatrix} \xi_\alpha & \bar\xi_{\dot\alpha} \end{pmatrix}. \tag{8.173}$$

Under Lorentz transformation $L_+^\uparrow$, Dirac spinor transforms as

$$\psi_D \to \psi_D'(x') = S(\Lambda)\psi_D(x)$$

$$= S(\Lambda)\psi_D(x'^\mu - \omega^{\mu\nu}x_\nu') \tag{8.174}$$

where

$$S(\Lambda) = 1 - \frac{i}{2}\omega^{\mu\nu}\frac{1}{2}\Sigma_{\mu\nu}. \tag{8.175}$$

Thus

$$\psi'_D(x) \equiv U\psi_D(x)U^\dagger = S(\Lambda)\left[\psi_D(x) - \omega^{\mu\nu}x_\nu\partial_\mu\psi_D(x)\right]$$

$$= S(\Lambda)\left[\psi_D - \frac{i}{2}\omega^{\mu\nu}L_{\mu\nu}\psi_D\right]. \tag{8.176}$$

Hence

$$[M^{\mu\nu}, \psi_D] = -J^{\mu\nu}\psi_D, \tag{8.177}$$

$$J^{\mu\nu} = L^{\mu\nu} + \frac{1}{2}\Sigma^{\mu\nu},$$

$$\frac{1}{2}\Sigma^{\mu\nu} = \frac{i}{4}\left[\gamma^\mu, \gamma^\nu\right]$$

$$= \begin{pmatrix} \sigma^{\mu\nu} & 0 \\ 0 & \bar{\sigma}^{\mu\nu} \end{pmatrix} \tag{8.178}$$

with

$$\sigma^{\mu\nu} = \frac{1}{4}\left(\sigma^\mu\bar{\sigma}^\nu - \sigma^\nu\bar{\sigma}^\mu\right),$$

$$\bar{\sigma}^{\mu\nu} = \frac{1}{4}\left(\bar{\sigma}^\mu\sigma^\nu - \bar{\sigma}^\nu\sigma^\mu\right). \tag{8.179}$$

Thus as far as spinor part is concerned

$$[M^{\mu\nu}, \psi] = -\frac{1}{2}\Sigma^{\mu\nu}\psi$$

$$= -i\begin{pmatrix} \sigma^{\mu\nu} & 0 \\ 0 & \bar{\sigma}^{\mu\nu} \end{pmatrix}\begin{pmatrix} \xi^\alpha \\ \bar{\chi}_{\dot{\alpha}} \end{pmatrix}. \tag{8.180}$$

Hence

$$[M^{\mu\nu}, \xi^\alpha] = -i\left(\sigma^{\mu\nu}\right)^\alpha{}_\beta\xi^\beta,$$

$$\left[M^{\mu\nu}, \bar{\chi}_{\dot{\alpha}}\right] = -i\left(\bar{\sigma}^{\mu\nu}\right)_{\dot{\alpha}}{}^{\dot{\beta}}\bar{\chi}_{\dot{\beta}}, \tag{8.181}$$

$$\left(\sigma_{\mu\nu}\right)^\alpha{}_\beta = \frac{1}{4}\left[\left(\sigma^\mu\right)^{\alpha\dot{\alpha}}\left(\bar{\sigma}^\nu\right)_{\dot{\alpha}\beta} - \mu \longleftrightarrow \nu\right],$$

$$\left(\bar{\sigma}^{\mu\nu}\right)_{\dot{\alpha}}{}^{\dot{\beta}} = \frac{1}{4}\left[\left(\bar{\sigma}^\mu\right)_{\dot{\alpha}\alpha}\left(\bar{\sigma}^\nu\right)^{\alpha\dot{\beta}} - \mu \longleftrightarrow \nu\right]. \tag{8.182}$$

8.9.2   *Spinor Charges*

Recall the Weyl spinors

$$\psi^\alpha = \begin{pmatrix} \xi^\alpha \\ \bar{\chi}_{\dot{\alpha}} \end{pmatrix}$$

$$\bar{\chi} = i\sigma^2 \chi^*$$

$$\chi = \xi^c$$

$$\bar{\chi}_{\dot{\alpha}} = i\sigma^2 \chi^{*\alpha} = \chi_\alpha^*. \tag{8.183}$$

Correspondingly, we introduce the spinor charge generators

$$Q^\alpha, \; \bar{Q}_{\dot{\alpha}} = i\sigma^2 Q^{*\alpha} = Q_\alpha^* \tag{8.184}$$

or equivalently 4-component Majorana spinor

$$Q_M \equiv \begin{pmatrix} Q^\alpha \\ \bar{Q}_{\dot{\alpha}} \end{pmatrix}. \tag{8.185}$$

We now discuss the commutation relations of the spinor charges with the Poincaré group generators. We first note that for two-component spinors, these are given in Eqs. (8.181) and (8.182).

Similarly for the spinor charges Q_α and $\bar{Q}_{\dot{\alpha}}$, we get

$$[M^{\mu\nu}, Q^\alpha] = -i(\sigma^{\mu\nu})^\alpha{}_\beta \, Q^\beta$$

$$\left[M^{\mu\nu}, \bar{Q}_{\dot{\alpha}} \right] = -i \, (\bar{\sigma}^{\mu\nu})_{\dot{\alpha}}{}^{\dot{\beta}} \, \bar{Q}_{\dot{\beta}}. \tag{8.186}$$

For the Poincaré generator P^μ

$$[P^\mu, Q^\alpha] = 0. \tag{8.187}$$

This is not obvious. To show this, we note that since P^μ is a 4-vector, only other 4-vector available is σ^μ. One would expect

$$[P^\mu, Q^\alpha] = c \, (\sigma^\mu)^{\alpha\dot{\beta}} \, \bar{Q}_{\dot{\beta}}. \tag{8.188}$$

Taking the complex conjugate $[P^\mu \sim -i\partial^\mu \to P^{*\mu} = -P^\mu]$

$$\left[P^\mu, (Q^\alpha)^* \right] = -c^* \, (\sigma^{*\mu})^{\alpha\dot{\beta}} \, (\bar{Q}_{\dot{\beta}})^*$$

or

$$\left[P^\mu, \bar{Q}^{\dot{\alpha}} \right] = -c^* \, (\bar{\sigma}^\mu)^{\dot{\alpha}\beta} \, Q_\beta$$

$$\epsilon^{\dot{\alpha}\dot{\gamma}} \left[P^\mu, \bar{Q}_{\dot{\gamma}} \right] = -c^* \, (\bar{\sigma}^\mu)^{\dot{\alpha}\beta} \, \epsilon_{\beta\alpha} Q^\alpha.$$

Multiplying by $\epsilon_{\dot\alpha\dot\beta}$ on both sides, we get

$$\epsilon_{\dot\alpha\dot\beta}\,\epsilon^{\dot\alpha\dot\gamma}\left[P^\mu,\bar{Q}_{\dot\gamma}\right]=-c^*\epsilon_{\dot\alpha\dot\beta}\,(\bar\sigma^\mu)^{\dot\alpha\beta}\,\epsilon_{\beta\alpha}Q^\alpha.$$

Thus

$$\left[P^\mu,\bar{Q}_{\dot\beta}\right]=-c^*(\bar\sigma^\mu)_{\dot\beta\alpha}\,Q^\alpha. \tag{8.189}$$

Then from the Jacobi identity

$$[P^\mu,[P^\nu,Q^\alpha]]+[P^\nu,[Q^\alpha,P^\mu]]+[Q^\alpha,[P^\mu,P^\nu]]=0$$

and using Eqs. (8.188) and (8.189), we get

$$|c|^2\left\{(\sigma^\mu)^{\alpha\dot\beta}\,(\bar\sigma^\nu)_{\dot\beta\gamma}-(\sigma^\nu)^{\alpha\dot\beta}\,(\bar\sigma^\mu)_{\dot\beta\gamma}\right\}Q^\gamma=0$$

or

$$|c|^2\,(\sigma^{\mu\nu})^\alpha_\gamma Q^\gamma=0$$

which implies $|c|^2=0$ or $c=0$. Hence we get

$$[P^\mu,Q^\alpha]=0. \tag{8.190}$$

To close the algebra we need to specify the anti-commutators $\{Q^\alpha,Q_\beta\}$, $\{Q^\alpha,\bar{Q}^{\dot\beta}\}$. Both of these are bosonic, rather than fermionic, so we require them to be linear in P^μ and $M^{\mu\nu}$, the generators of the Poincaré group.

Thus

$$\{Q^\alpha,\bar{Q}^{\dot\beta}\}=t\,(\sigma^\mu)^{\alpha\dot\beta}\,P_\mu,$$

$$\{Q^\alpha,Q_\beta\}=s\,(\sigma^{\mu\nu})^\alpha_\beta\,M_{\mu\nu}. \tag{8.191}$$

Now $\{Q^\alpha,\bar{Q}^{\dot\beta}\}$ transforms as $(1/2,1/2)$ under $L_+^\uparrow$ and P^μ. Since Q^α,Q^β and $\bar{Q}^{\dot\beta}$ all commute with P_λ, we have

$$0=[P_\lambda,\{Q^\alpha,Q_\beta\}]=s\,(\sigma^{\mu\nu})^\alpha_{\ \beta}\,[P_\lambda,M_{\mu\nu}] \tag{8.192}$$

where

$$[P_\lambda,M_{\mu\nu}]=-i\,(\eta_{\nu\lambda}P_\mu-\eta_{\mu\lambda}P_\nu) \tag{8.193}$$

implying $s=0$. Now

$$\left[P_\lambda,\left\{Q^\alpha,\bar{Q}^{\dot\beta}\right\}\right]=t\left[P_\lambda,(\sigma^\mu)^{\alpha\dot\beta}\,P_\mu\right]$$

$$=t\,(\sigma^\mu)^{\alpha\dot\beta}\,[P_\lambda,P_\mu]$$

$$=0 \tag{8.194}$$

since the left-hand side is zero, it is an identity.

Thus

$$\{Q^\alpha, Q^\beta\} = 0 = \{\bar{Q}_{\dot\alpha}, \bar{Q}_{\dot\beta}\}. \tag{8.195}$$

We choose $t = 2$ as the normalization condition so that

$$\{Q^\alpha, \bar{Q}^{\dot\beta}\} = 2\,(\sigma^\mu)^{\alpha\dot\beta}\,P_\mu. \tag{8.196}$$

To summarize: Spinor Algebra

$$[M^{\mu\nu}, Q^\alpha] = -i\,(\sigma^{\mu\nu})^\alpha{}_\beta Q^\beta \tag{8.197}$$

$$\left[M^{\mu\nu}, \bar{Q}_{\dot\alpha}\right] = -i\,(\bar{\sigma}^{\mu\nu})_{\dot\alpha}{}^{\dot\beta}\,\bar{Q}_{\dot\beta} \tag{8.198}$$

$$[P^\mu, Q^\alpha] = 0 = \left[P^\mu, Q^{\dot\alpha}\right] \tag{8.199}$$

$$\{Q^\alpha, Q^\beta\} = 0 = \left\{Q^{\dot\alpha}, Q^{\dot\beta}\right\} \tag{8.200}$$

$$\{Q^\alpha, \bar{Q}^{\dot\beta}\} = 2\,(\sigma^\mu)^{\alpha\dot\beta}\,P_\mu = 2\,(\bar{\sigma}^\mu)^{\dot\beta\alpha}\,P_\mu. \tag{8.201}$$

In order to discuss the consequence of the SUSY algebra, we note

$$\bar{\sigma}^{\mu\nu} = \frac{1}{4}\left(\bar{\sigma}^\mu\sigma^\nu - \bar{\sigma}^\nu\sigma^\mu\right),$$

$$\sigma^{\mu\nu} = \frac{1}{4}\left(\sigma^\mu\bar{\sigma}^\nu - \sigma^\nu\bar{\sigma}^\mu\right),$$

$$\sigma^\mu\bar{\sigma}^\nu + \sigma^\nu\bar{\sigma}^\mu = 2\eta^{\mu\nu}. \tag{8.202}$$

Thus

$$Tr\,(\sigma^\mu\bar{\sigma}^\nu) = 2\eta^{\mu\nu}. \tag{8.203}$$

From Eq. (8.201), one gets (using Eq. (8.203))

$$(\sigma^\mu)_{\alpha\dot\beta}\left\{Q^\alpha, \bar{Q}^{\dot\beta}\right\} = 2\,(\sigma^\mu)_{\alpha\dot\beta}\,(\bar{\sigma}^\nu)^{\dot\beta\alpha}\,P_\nu$$

$$= 2Tr\,(\sigma^\mu\bar{\sigma}^\nu)\,P_\nu$$

$$= 4P^\mu. \tag{8.204}$$

For $\mu = 0$

$$(\sigma^0)_{\alpha\dot\beta}\left\{Q^\alpha, \bar{Q}^{\dot\beta}\right\} = 4P^0. \tag{8.205}$$

Since

$$\sigma^0 = \begin{pmatrix} 1 & 0 \\ 0 & 1 \end{pmatrix},$$

from Eq. (8.205)

$$\delta_{\alpha\dot\beta}\left\{Q^\alpha,\bar Q^{\dot\beta}\right\} = \delta_{\alpha\dot\beta}\left(Q^\alpha\bar Q^{\dot\beta} + \bar Q^{\dot\beta}Q^\alpha\right) = 4P^0$$

therefore,

$$Q^1\bar Q^{\dot1} + Q^2\bar Q^{\dot2} + \bar Q^{\dot1}Q^1 + \bar Q^{\dot2}Q^2 = 4P^0. \tag{8.206}$$

Hence

$$4\left\langle\Psi\left|P^0\right|\Psi\right\rangle = \sum_{\alpha=1}^{2}\left\langle\Psi\left|Q^\alpha(Q^\alpha)^* + (Q^\alpha)^*Q^\alpha\right|\Psi\right\rangle \geq 0 \tag{8.207}$$

implying that the spectrum of $P^0 = H$ is semi-positive definite. In particular for vacuum state $|0\rangle$, $E_{vac} = 0$ implies

$$\langle0\left|P_0\right|0\rangle = 0 \implies Q^\alpha|0\rangle = 0. \tag{8.208}$$

Thus the vanishing of vacuum energy is a necessary and sufficient condition for the existence of a unique vacuum.

The two invariants of the Poincaré groups are $P^\mu P_\mu$ and $W^2 = W_\mu W^\mu$, where

$$W^\mu = -\frac{1}{2}\varepsilon^{\mu\nu\rho\sigma}M_{\nu\rho}P_\sigma. \tag{8.209}$$

Since $[M^{\mu\nu}, Q^\alpha] \neq 0$, it follows that

$$[P^\mu, Q^\alpha] = 0 = \left[P^\mu, Q^{\dot\alpha}\right],$$

$$\left[W^2, Q^\alpha\right] \neq 0, \tag{8.210}$$

i.e., W^2 is not invariant for SUSY algebra.

For particle of mass m at rest, $W^\mu = \frac{m}{2}\varepsilon^{0\mu\rho\sigma}M_{\nu\rho}$, therefore $W^0 = 0$ and

$$W^i = \frac{m}{2}\epsilon^{ijk}M_{jk} = mJ^i, \tag{8.211}$$

$$W^2 = -m^2\mathbf{J}^2.$$

Now

$$[W^i, Q^\alpha] = \frac{m}{2}\epsilon^{ijk}[M_{jk}, Q^\alpha]$$

$$= -\frac{m}{2}(\sigma^i)^\alpha_\beta\, Q^\beta$$

or

$$[J^i, Q^\alpha] = -\frac{1}{2}(\sigma^i)^\alpha_\beta\, Q^\beta. \tag{8.212}$$

Hence it follows that for massive particles, the irreducible representations of the SUSY algebra will contain different spins. From Eq. (8.212), it is clear that $j = 1/2$ super-charge Q^α acting on a state of spin j results in a state of spin $j \pm 1/2$, thereby mixing fermions and bosons.

Since Q^α changes fermion number by one unit, we may write

$$(-1)^{N_F} Q^\alpha = (-1)^{N_F+1} Q^\alpha = -(-1)^{N_F} Q^\alpha, \qquad (8.213)$$

where N_F is the fermion number operator.

Now

$$Tr \left[(-1)^{N_F} \left\{ Q^\alpha, \bar{Q}^{\dot{\beta}} \right\} \right]$$

$$= Tr \left[-Q^\alpha (-1)^{N_F} \bar{Q}^{\dot{\beta}} + Q^\alpha (-1)^{N_F} \bar{Q}^{\dot{\beta}} \right] = 0. \qquad (8.214)$$

Thus using Eq. (8.196), we get from Eq. (8.214)

$$Tr \left[(-1)^{N_F} 2 (\sigma^\mu)^{\alpha \dot{\beta}} P_\mu \right] = 0. \qquad (8.215)$$

Hence, for a fixed non-zero P_μ,

$$Tr \left[(-1)^{N_F} \right] \equiv \sum_m \left\langle m \left| (-1)^{N_F} \right| m \right\rangle = 0. \qquad (8.216)$$

Since $(-1)^{N_F}$ has value $+1$ on a bosonic state and -1 on a fermionic state, this implies that for a finite dimensional representation R of SUSY algebra

$$n_B (R) - n_F (R) = 0$$

where n_B is the number of bosons in the representation R and n_F is the number of fermions in the representation R.

8.10   Supersymmetric Multiplets

In this section, the representations of the SUSY algebra that can be realized for massless one particle states are considered. The massless case is most interesting, as in most of the phenomenologically interesting scenarios non-zero masses of particles that we observed are generated by SUSY breaking.

For massless particle

$$P^2 = 0, \ \mathbf{W}^2 = -m^2 \mathbf{J}^2 = 0,$$

$$W^\lambda P_\lambda = 0, \ \implies W^\mu = \lambda P^\mu, \qquad (8.217)$$

where

$$\lambda = \frac{\mathbf{J} \cdot \mathbf{P}}{P_0} \tag{8.218}$$

is the Helicity.

Now $W^0 = \mathbf{J} \cdot \mathbf{P}$, thus from Eqs. (8.197) and (8.198):

$$\left[W^0, Q^\alpha\right] = \left[\mathbf{J} \cdot \mathbf{P}, Q^\alpha\right] = -\frac{1}{2}(\sigma \cdot \mathbf{P})^\alpha{}_\beta Q^\beta,$$

$$\left[W^0, \bar{Q}_{\dot\alpha}\right] = -\frac{1}{2}(\sigma \cdot \mathbf{P})_{\dot\alpha}{}^{\dot\beta} \bar{Q}_{\dot\beta}. \tag{8.219}$$

For a massless state $|p, \lambda\rangle$:

$$P^\mu |p, \lambda\rangle = p^\mu |p, \lambda\rangle, \tag{8.220}$$

$$W^\mu |p, \lambda\rangle = \lambda p^\mu |p, \lambda\rangle. \tag{8.221}$$

Select

$$P^\mu = (P, 0, 0, P), \ p^\mu = (E, 0, 0, E) \ E > 0. \tag{8.222}$$

Thus from

$$\left\{Q^\alpha, \bar{Q}^{\dot\beta}\right\} = 2\,(\sigma^\mu)^{\alpha\dot\beta} P_\mu$$

or equivalently

$$\left\{Q^\alpha, (Q^*)^\beta\right\} = 2\,(\sigma^\mu)^{\alpha\beta} P_\mu, \tag{8.223}$$

we have

$$\left\{Q^\alpha, (Q^*)^\beta\right\} = 2\delta^{\alpha\beta} P_0 + 2\,(\sigma^3)^{\alpha\beta} P_3. \tag{8.224}$$

Hence from Eq. (8.224):

$$\left\{Q^1, (Q^1)^*\right\} = 2\,(P^0 - P^3) = 0$$

$$\left\{Q^2, (Q^2)^*\right\} = 2\,(P^0 + P^3) = 2P. \tag{8.225}$$

Equation (8.225) implies

$$\left\langle \Psi \left| \left\{Q^1, (Q^1)^*\right\} \right| \Psi \right\rangle = 0$$

or

$$\|Q^{1*}\Psi\|^2 + \|Q^1\Psi\|^2 = 0,$$

so that

$$Q^1 |\Psi\rangle = 0, \ \bar{Q}^{\dot 1} |\Psi\rangle = 0,$$

but

$$Q^2|\Psi\rangle \neq 0, \ \bar{Q}^2|\Psi\rangle \neq 0. \tag{8.226}$$

From Eq. (8.219):

$$\left[W^0, Q^1\right] = -\frac{1}{2}PQ^1,$$

$$\left[W^0, Q^2\right] = \frac{1}{2}PQ^2, \tag{8.227}$$

$$\left[W^0, \bar{Q}_{\dot{1}}\right] = -\frac{1}{2}P\bar{Q}_{\dot{1}},$$

$$\left[W^0, \bar{Q}_{\dot{2}}\right] = \frac{1}{2}P\bar{Q}_{\dot{2}}. \tag{8.228}$$

Now

$$\bar{Q}_{\dot{\alpha}} = \varepsilon_{\dot{\alpha}\dot{\beta}}\bar{Q}^{\dot{\beta}} \implies \bar{Q}_{\dot{1}} = \bar{Q}^{\dot{2}}, \ \bar{Q}_{\dot{2}} = -\bar{Q}^{\dot{1}},$$

therefore, Eq. (8.228) can be rewritten as

$$\left[W^0, \bar{Q}^{\dot{2}}\right] = -\frac{1}{2}P\bar{Q}^{\dot{2}},$$

$$\left[W^0, \bar{Q}^{\dot{1}}\right] = \frac{1}{2}P\bar{Q}^{\dot{1}}. \tag{8.229}$$

Now

$$W^0|p,\lambda\rangle = \mathbf{J} \cdot \mathbf{P}|p,\lambda\rangle$$

$$= PJ^3|p,\lambda\rangle = E\lambda|p,\lambda\rangle. \tag{8.230}$$

Thus from Eq. (8.229):

$$W^0\bar{Q}^{\dot{2}}|p,\lambda\rangle = \left[-\frac{1}{2}P\bar{Q}^{\dot{2}} + \bar{Q}^{\dot{2}}W^0\right]|p,\lambda\rangle = E\left(\lambda - \frac{1}{2}\right)\bar{Q}^{\dot{2}}|p,\lambda\rangle \tag{8.231}$$

and from Eq. (8.227):

$$W^0Q^2|p,\lambda - \frac{1}{2}\rangle = E\lambda Q^2|p,\lambda\rangle. \tag{8.232}$$

Thus if $\bar{Q}^{\dot{2}}|p,\lambda\rangle$ is not zero, it is an eigenstate of W^0 with eigenvalue $E\left(\lambda - \frac{1}{2}\right)$. Similarly, if $Q^2|p,\lambda\rangle \neq 0$, it is an eigenstate of W^0 with eigenvalue $E\left(\lambda + \frac{1}{2}\right)$.

To summarize

$$Q^1|p,\lambda,R\rangle = 0, \ Q^{\dot{1}}|p,\lambda,R\rangle = 0, \tag{8.233}$$

$$\bar{Q}^{\dot{2}}|p,\lambda,R\rangle = N|p,\lambda - \frac{1}{2},R\rangle = \sqrt{4E}|p,\lambda - \frac{1}{2},R\rangle, \tag{8.234}$$

$$Q^2|p,\lambda - \frac{1}{2},R\rangle = \sqrt{4E}|p,\lambda,R\rangle. \tag{8.235}$$

Equivalently

$$\bar{Q}_{\dot{1}}|p,\lambda,R\rangle = \sqrt{4E}|p,\lambda - \frac{1}{2},R\rangle, \qquad (8.236)$$

$$Q_1|p,\lambda - \frac{1}{2},R\rangle = \sqrt{4E}|p,\lambda,R\rangle. \qquad (8.237)$$

R specify the other quantum numbers.

$$CTP\,|p,\lambda,R\rangle = |p,-\lambda,R^*\rangle\,.$$

Hence there are just two states with helicity λ and $\lambda - \frac{1}{2}$. The most common of them, which we encounter are

$$\lambda = \frac{1}{2},\ 1,\ 2.$$

$\lambda = \frac{1}{2}$: Chiral Multiplet: a Weyl spinor with helicity $\frac{1}{2}$ and a scalar, plus CPT conjugate Weyl fermion of helicity $-\frac{1}{2}$ and another scalar.

$\lambda = 1$: Vector Supermultiplet: a massless vector particle, a fermion of helicity $\frac{1}{2}$ and CPT conjugate $\lambda = -1$ and $\lambda = -\frac{1}{2}$.

$\lambda = 2$: Graviton Supermultiplet $\left(2,\frac{3}{2}\right)$ and CPT conjugate $\lambda = -2$, $\lambda = -\frac{3}{2}$.

To conclude: Bosonic particles are naturally paired with fermionic ones. Each minimal pairing is called supermultiplet: As we have seen above, a left-handed fermion $\left(\lambda = \frac{1}{2}\right)$; its right-handed antiparticle $\left(\lambda = -\frac{1}{2}\right)$, a complex boson $(\lambda = 0)$ and its conjugate form a supermultiplet. A massless vector field $(\lambda = 1)$ and a left-handed fermion form a vector multiplet, two transversely polarized vector boson states, plus left-handed and right-handed fermion and its anti-particle. Thus in $N = 1$ supersymmetry one has the helicity states given in the Table below:

Particle	λ	Helicity	Degeneracy	CPT conjugate helicity	Supermultiplet
Quark, lepton	1/2	1/2	1	$-1/2$	Chiral
Higgsino		1/2	1		
squark, slepton, Higgs		0	1	0	
Gauge boson	1	1	1	-1	vector
Gaugino		1/2	1	$-1/2$	
Graviton	2	2	1	-2	Gravity
Gravitino		3/2	1	$-1/2$	

8.11 Problems

8.1 Show that

$$\mathbf{W} = P^0 \mathbf{J} - \mathbf{P} \times \mathbf{K}.$$

8.2 Show that

$$\left[K^i, P^0\right] = -iP^i,$$
$$\left[K^i, P^j\right] = -i\delta^{ij} P^0.$$

8.3 For a massless particle, show that

$$[W^1, W^2] = 0,$$
$$[\frac{W^0}{P^0}, W^1] = iW^2,$$
$$[\frac{W^0}{P^0}, W^2] = iW^1.$$

8.4 Show that
   a)

$$[AB, CD] = \{A, C\}DB - AC\{B, D\} + A\{C, B\}D - C\{A, D\}B.$$

b) If the matrices Γ_a, Γ_b $(a, b = 1, 2, 3, 4)$ satisfy the anticommutation relation

$$\{\Gamma_a, \Gamma_b\} = 2\delta_{ab}.$$

Using (a), show that matrices

$$R_{ab} = -\frac{i}{4}(\Gamma_a \Gamma_b - \Gamma_b \Gamma_a)$$

satisfy the commutation relation

$$[R_{ab}, R_{cd}] = i[\delta_{ac} R_{db} - \delta_{bd} R_{ac} + \delta_{cb} R_{ad} - \delta_{ad} R_{cb}],$$

i.e., generate the algebra of rotation group $O(4)$.

8.5 Show that P^2 and W^2 commute write all the generators of the Poincaré group.

8.6 Verify

$$[W^\lambda, M^{\rho\sigma}] = -i[W^\rho M^{\lambda\sigma} - W^\sigma M^{\lambda\rho}]$$

using the Lie algebra of Poincaré group.

8.7 Show that

$$[W^\mu, W^\nu] = i\epsilon^{\mu\nu\rho\sigma} W_\rho P_\sigma.$$

Solution

$$[W^\mu, W^\nu] = -\frac{1}{2}\epsilon^{\mu\alpha\beta\rho} [M_{\alpha\beta} P_\rho, W^\nu]$$

$$= -\frac{1}{2}\epsilon^{\mu\alpha\beta\rho} \{M_{\alpha\beta} [P_\rho, W^\nu] + [M_{\alpha\beta}, W^\nu] P_\rho\}$$

$$= -\frac{1}{2}\epsilon^{\mu\alpha\beta\rho} \{g^{\nu\tau} [M_{\alpha\beta}, W_\tau] P_\rho\}$$

$$= -\frac{1}{2}\epsilon^{\mu\alpha\beta\rho} i g^{\nu\tau} (g_{\beta\tau} W_\alpha - g_{\alpha\tau} W_\beta) P_\rho$$

$$= -\frac{i}{2}\epsilon^{\mu\alpha\beta\rho} (\delta^\nu_\beta W_\alpha - \delta^\nu_\alpha W_\beta) P_\rho$$

$$= i\epsilon^{\mu\nu\lambda\rho} W_\lambda P_\rho.$$

8.8 Show that $\mathbf{J}^2 - \mathbf{K}^2$ and $\mathbf{J} \cdot \mathbf{K}$ commute with all the six generators of L^+ group. Show further that

$$J_{\mu\nu} J^{\mu\nu} = 2(\mathbf{J}^2 - \mathbf{K}^2)$$

$$\epsilon^{\mu\nu\rho\sigma} J_{\mu\nu} J_{\rho\sigma} = -\mathbf{J} \cdot \mathbf{K}.$$

8.9 Consider special conformal transformation

$$x^\mu \longrightarrow x'^\mu = \frac{x^\mu - c^\mu x^2}{\sigma(x)} ; \sigma(x) = 1 - 2c^\mu x_\mu + c^2 x^2.$$

Using the above transformation

$$(x - y)^2 \longrightarrow (x' - y')^2 = \frac{(x - y)^2}{\sigma(x)\sigma(y)}.$$

Infinitesimal Transformation; put $c_\mu = -\epsilon_\mu$

$$x'^\mu = x^\mu + \epsilon_\nu(\eta^{\mu\nu} x^2 - 2x^\nu x^\mu)$$

$$= \frac{x^\mu}{2}(1 - 2\epsilon.x) + \frac{1}{2} (x^\mu + \epsilon^{\mu\nu} x_\nu)$$

$$\epsilon^{\mu\nu} = 2 (\epsilon^\mu x^\nu - \epsilon^\nu x^\mu).$$

Written in this form: Conformal Transformation $\sim$ Scale Transformation + Lorentz Transformation.

Corresponding unitary operator is

$$U = 1 + i\epsilon^\mu K_\mu$$

$$U\phi(x)U^+ = \phi'(x) = (1 + \epsilon^\mu t_\mu)\phi(x^\mu - \epsilon_\nu(\eta^{\mu\nu}x^2 - 2x^\nu x^\mu)).$$

Show that

$$[K_\mu, \phi(x)] = [t_\mu + \overline{k}_\mu]\phi(x),$$

where

$$\overline{k}_\mu = i(x^2\partial_\mu - 2x_\mu x^\lambda\partial_\lambda) = i[-x_\mu x^\nu\partial_\nu + x^\nu(x_\nu\partial_\mu - \partial_\nu x_\mu)]$$

$$= -ix_\mu x^\nu\partial_\nu + x^\nu L_{\nu\mu}.$$

Noting that conformal transformation is a combination of homogeneous Lorentz Transformation and scale transformation

$$t_\mu = -ilx_\mu + x^\nu S_{\nu\mu}.$$

Hence

$$[K_\mu, \phi(x)] = [-ix_\mu(l + x^\nu\partial_\nu) + x^\nu J_{\nu\mu}]\phi(x).$$

8.10 Under Lorentz transformations, the spinors ξ and $\overline{\xi}$ transform as

$$\xi' = e^{\frac{1}{2}(i\omega\cdot\sigma + \varsigma\cdot\sigma)}\xi,$$

$$\overline{\xi}' = e^{\frac{1}{2}(i\omega\cdot\sigma - \varsigma\cdot\sigma)}\overline{\xi},$$

$$\omega = \theta\mathbf{n}, \quad \varsigma = \varsigma\mathbf{e}_v, \quad \mathbf{e}_v = \frac{\mathbf{v}}{v}.$$

How do the vectors

(i)

$$A^\mu = \xi^\dagger\sigma^\mu\xi$$

(ii)

$$B^\mu = \xi^\dagger\sigma^\mu\overline{\xi}$$

transform under rotation and Lorentz boost?

Solution

$$e^{\frac{1}{2}i\omega\cdot\sigma} = e^{\frac{i}{2}\theta\mathbf{n}\cdot\sigma} = \cos\frac{\theta}{2} + i\mathbf{n}\cdot\sigma\sin\frac{\theta}{2},$$

$$e^{\frac{1}{2}\varsigma\mathbf{e}_v\cdot\sigma} = 1 + \frac{\xi}{2}(\mathbf{e}_v\cdot\sigma) + \frac{1}{2!}\left(\frac{\xi}{2}\right)^2(\mathbf{e}_v\cdot\sigma)^2 + \frac{1}{3!}\left(\frac{\xi}{2}\right)^3(\mathbf{e}_v\cdot\sigma)^3 + \cdots$$

$$= \left(1 + \frac{1}{2!}\left(\frac{\xi}{2}\right)^2 + \cdots\right) + (\mathbf{e}_v\cdot\sigma)\left(\frac{\xi}{2} + \left(\frac{\xi}{2}\right)^3 + \cdots\right)$$

$$= \cosh\frac{\xi}{2} + (\mathbf{e}_v\cdot\sigma)\sinh\frac{\xi}{2}.$$

(1) (i) Under rotation

$$A^i = A'^i = \xi^{\dagger\prime}\sigma^i\xi'$$

$$= \xi^\dagger \left[\left(\cos\frac{\theta}{2} - i\mathbf{n}\cdot\sigma\sin\frac{\theta}{2} \right) \sigma^i \left(\cos\frac{\theta}{2} + i\mathbf{n}\cdot\sigma\sin\frac{\theta}{2} \right) \right] \xi.$$

Using

$$\sigma^i\sigma^j + \sigma^j\sigma^i = 2\delta^{ij}$$

$$\sigma^i\sigma^j - \sigma^j\sigma^i = 2i\epsilon^{ijk}\sigma^k \tag{8.238}$$

$$\sigma^j\sigma^i\sigma^k = i\epsilon^{ijk} + 2\delta^{ji}\sigma^k - \delta^{jk}\sigma^i$$

we have

$$A'^i = \xi^\dagger \left[\left(\cos^2\frac{\theta}{2} - \sin^2\frac{\theta}{2} \right) \sigma^i - 2\sin\frac{\theta}{2}\cos\frac{\theta}{2} \left(\mathbf{n}\times\sigma \right)_i + \sin^2\frac{\theta}{2} \left(2\mathbf{n}\cdot\sigma n^i \right) \right] \xi$$

or

$$\mathbf{A}' = \cos\theta\mathbf{A} - \sin\theta\mathbf{n}\times\mathbf{A} + (1 - \cos\theta)\left(\mathbf{n}\cdot\mathbf{A} \right)\mathbf{n}$$

$$A'^0 = A^0.$$

Similarly for the B^μ.

Under Lorentz boost

$$A'^i = \xi^\dagger \left[\left(\cosh\frac{\varsigma}{2} + \mathbf{e}_v\cdot\sigma\sinh\frac{\varsigma}{2} \right) \sigma^i \left(\cosh\frac{\varsigma}{2} + \mathbf{e}_v\cdot\sigma\sinh\frac{\varsigma}{2} \right) \right] \xi.$$

Using Eq. (8.238)

$$\mathbf{A}' = \mathbf{A} + \sinh\varsigma\frac{\mathbf{v}}{v}A^0 + (\cosh\varsigma - 1)\frac{\mathbf{v}}{v}\frac{\mathbf{v}\cdot\mathbf{A}}{v},$$

$$A'^0 = \xi'^\dagger\xi' = \xi^\dagger \left[\left(\cosh^2\frac{\varsigma}{2} + \sinh^2\frac{\varsigma}{2} \right) + \cosh\frac{\varsigma}{2}\sinh\frac{\varsigma}{2}\mathbf{e}_v\cdot\sigma \right] \xi$$

$$= \cosh\varsigma A^0 + \sinh\varsigma\frac{\mathbf{v}\cdot\mathbf{A}}{v}.$$

Now

$$\cosh\varsigma = \gamma, \quad \sinh\varsigma = \gamma\frac{v}{c}.$$

Hence

$$\mathbf{A}' = \mathbf{A} - (1 - \gamma)\frac{\mathbf{v}\cdot\mathbf{A}}{v^2}\mathbf{v} + \gamma\frac{v}{c}A^0$$

$$A'^0 = \gamma \left(A^0 + \frac{\mathbf{v}\cdot\mathbf{A}}{c} \right)$$

(ii)

$$B^\mu = \xi^\dagger\sigma^\mu\bar{\xi}.$$

Under rotation $\mathbf{B}$ and B^0 transforms as $\mathbf{A}$ and A^0. However, under Lorentz boost

$$B'^i = \xi^{\dagger'}\sigma^i\bar{\xi}'$$

$$= \xi^{\dagger}\left[\left(\cosh\frac{\varsigma}{2} + \mathbf{e}_v\cdot\sigma\sinh\frac{\varsigma}{2}\right)\sigma^i\left(\cosh\frac{\varsigma}{2} - \mathbf{e}_v\cdot\sigma\sinh\frac{\varsigma}{2}\right)\right]\bar{\xi}$$

$$= \xi^{\dagger}\left[\cosh^2\frac{\varsigma}{2}\sigma^i - \sinh\frac{\varsigma}{2}\cosh\frac{\varsigma}{2}\left(\sigma^i\mathbf{e}_v\cdot\sigma - \mathbf{e}_v\cdot\sigma\sigma^i\right) - \sinh^2\frac{\varsigma}{2}\left(\mathbf{e}_v\cdot\sigma\sigma^i\mathbf{e}_v\cdot\sigma\right)\right]\bar{\xi}.$$

Thus

$$\mathbf{B}' = \left[\cosh\varsigma\mathbf{B} - i\sinh\varsigma\,(\mathbf{e}_v\times\mathbf{B}) + (1-\cosh\varsigma)\frac{\mathbf{v}\cdot\mathbf{B}}{v^2}\mathbf{v}\right],$$

$$B'^0 = \xi^{\dagger'}\bar{\xi}'$$

$$= \xi^{\dagger}\left[\left(\cosh\frac{\varsigma}{2} + \mathbf{e}_v\cdot\sigma\sinh\frac{\varsigma}{2}\right)\left(\cosh\frac{\varsigma}{2} - \mathbf{e}_v\cdot\sigma\sinh\frac{\varsigma}{2}\right)\right]\bar{\xi}$$

$$= \xi^{\dagger}\bar{\xi} = B^0.$$

Hence

$$\mathbf{B}' = \gamma\left[\mathbf{B} - i\frac{\mathbf{v}\times\mathbf{B}}{c} + \frac{1-\gamma}{\gamma}\frac{\mathbf{v}\cdot\mathbf{B}}{v^2}\mathbf{v}\right],$$

$$B'^0 = B^0.$$

8.11 Consider the SUSY Lagrangian

$$\mathcal{L} = \frac{1}{2}\dot{q}^2 - \frac{1}{2}[V(q)]^2 = +\frac{i}{2}(\bar{\psi}\dot{\psi} - \dot{\bar{\psi}}\psi) - V^2(q)\bar{\psi}\psi$$

(i) Write down the Lagrange equations of motion for q, ψ and $\bar{\psi}$.

(ii) Using equation of motions show that

$$\delta\mathcal{L} = \frac{d}{dt}[\dot{q}\delta q + \frac{i}{2}(\bar{\psi}\delta\psi - \delta\bar{\psi}\psi)].$$

(iii) Show that the invariance of the Lagrangian under the transformation

$$\delta q = \xi\psi + \bar{\psi}\bar{\xi}$$

$$\delta\psi = (i\dot{q} - V(q))\bar{\xi}$$

$$\delta\bar{\psi} = -(i\dot{q} - V(q))\xi$$

where ξ are anticommuting (constant) parameter, lead to conserved charges

$$Q = (-ip + V(q))\psi,$$

$$Q^+ = \bar{\psi}(ip + V(q)).$$

Hint: the $\xi\psi = -\psi\xi$ in the last step.

Solution

Parts (i) and (iii) are already covered in the text. The solution of part (ii) is given below.

$$\delta\mathcal{L} = \frac{\partial\mathcal{L}}{\partial\dot{q}}\delta\dot{q} + \frac{\partial\mathcal{L}}{\partial q}\delta q + \delta\dot{\overline{\psi}}\frac{\partial\mathcal{L}}{\partial\dot{\overline{\psi}}} + \delta\dot{\psi}\frac{\partial\mathcal{L}}{\partial\dot{\overline{\psi}}} + \frac{\partial\mathcal{L}}{\partial\psi}\delta\psi + \frac{\partial\mathcal{L}}{\partial\overline{\psi}}\delta\overline{\psi},$$

$$\frac{\partial\mathcal{L}}{\partial\dot{q}} = \dot{q}, \quad \frac{\partial\mathcal{L}}{\partial q} = -V(q)\,V'(q) - V''(q)\,\overline{\psi}\psi$$

$$\frac{\partial\mathcal{L}}{\partial\dot{\psi}} = \frac{i}{2}\overline{\psi}, \quad \frac{\partial\mathcal{L}}{\partial\dot{\overline{\psi}}} = -\frac{i}{2}\psi,$$

$$\frac{\partial\mathcal{L}}{\partial\psi} = -\frac{i}{2}\dot{\overline{\psi}} - V'(q)\,\overline{\psi}, \quad \frac{\partial\mathcal{L}}{\partial\overline{\psi}} = \frac{i}{2}\dot{\psi} - V'(q)\,\psi.$$

Therefore

$$\delta\mathcal{L} = \dot{q}\delta\dot{q} - \left(V(q)\,V'(q) + V''(q)\,\overline{\psi}\psi\right)\delta q - \frac{i}{2}\delta\overline{\psi}\dot{\psi} + \frac{i}{2}\overline{\psi}\delta\dot{\psi}$$

$$+\delta\overline{\psi}\left(\frac{i}{2}\dot{\psi} - V'(q)\,\psi\right) + \left(-\frac{i}{2}\dot{\overline{\psi}} - V'(q)\,\overline{\psi}\right)\delta\psi.$$

Hence, using the equations of motion

$$\delta\mathcal{L} = \dot{q}\delta\dot{q} - \ddot{q}\delta q - \frac{i}{2}\delta\overline{\psi}\dot{\psi} + \frac{i}{2}\overline{\psi}\delta\dot{\psi} - \frac{i}{2}\delta\dot{\overline{\psi}}\psi + \frac{i}{2}\dot{\overline{\psi}}\delta\psi$$

$$= \frac{d}{dt}\left[\dot{q}\delta q + \frac{i}{2}\left(\overline{\psi}\delta\psi - \delta\overline{\psi}\psi\right)\right].$$

8.12 Show that

$$(C\gamma^{\mu\nu})^T = C\gamma^{\mu\nu},$$

$$(C\gamma^5\gamma^\mu)^T = -C\gamma^5\gamma^\mu,$$

$$\gamma^{\mu\nu} = [\gamma^\mu, \gamma^\nu].$$

8.13 Show that

$$\overline{\Psi}_D\gamma^\mu\Psi_D = \chi_\alpha\,(\sigma^\mu)^{\alpha\dot{\beta}}\,\overline{\chi}_{\dot{\beta}} - \xi_\alpha\,(\sigma^\mu)^{\alpha\dot{\beta}}\,\overline{\xi}_{\dot{\beta}}$$

and

$$\overline{\Psi}_M\gamma^\mu\Psi_M = 0.$$

8.14 Show that

$$\sigma^2 \sigma^\mu \sigma^2 = (\overline{\sigma}^\mu)^T,$$

$$(\sigma^\mu)_{\alpha\dot{\beta}} = (\overline{\sigma}^\mu)_{\dot{\beta}\alpha},$$

$$\overline{\chi}^{\dot{\alpha}} (\overline{\sigma}^\mu)_{\dot{\alpha}\beta} \xi^\beta = \overline{\chi}_{\dot{\alpha}} (\overline{\sigma}^\mu)^{\dot{\alpha}\beta} \xi_\beta = -\xi_\beta (\sigma^\mu)^{\beta\dot{\alpha}} \overline{\chi}_{\dot{\alpha}},$$

$$\chi_\alpha (\sigma^{\mu\nu})^\alpha {}_\beta \xi^\beta = \chi^\alpha (\sigma^{\mu\nu})_\alpha {}^\beta \xi_\beta = -\xi_\beta (\sigma^{\mu\nu})^\beta {}_\alpha \chi^\alpha.$$

8.15 Show that

$$[W^\mu, Q^\alpha] = \frac{i}{2} \varepsilon^{\mu\nu\rho\gamma} P_\nu (\sigma_{\rho\gamma})^\alpha {}_\beta Q^\beta,$$

$$[W^\mu, \overline{Q}_{\dot{\alpha}}] = \frac{i}{2} \varepsilon^{\mu\nu\rho\gamma} P_\nu (\sigma_{\rho\gamma})_{\dot{\alpha}} {}^{\dot{\beta}} \overline{Q}_{\dot{\beta}}.$$

Then show that

$$[W^0, Q^\alpha] = -\frac{1}{2} (\sigma \cdot \mathbf{P})^\alpha {}_\beta Q^\beta,$$

$$[W^0, \overline{Q}_{\dot{\alpha}}] = -\frac{1}{2} (\sigma \cdot \mathbf{P})_{\dot{\alpha}} {}^{\dot{\beta}} \overline{Q}_{\dot{\beta}}.$$

8.16 Show that for a massive particle of mass m at rest

$$[W^i, Q^\alpha] = -\frac{m}{2} (\sigma^i)^\alpha {}_\beta Q^\beta,$$

$$[W^i, \overline{Q}_{\dot{\alpha}}] = -\frac{m}{2} (\sigma^i)_{\dot{\alpha}} {}^{\dot{\beta}} \overline{Q}_{\dot{\beta}}.$$

Using $W^i = mJ^i$. Selecting,

$$Q^1 |j, j_3\rangle = 0 \text{ and } Q^2 |j, j_3\rangle = 0$$

show that

$$\frac{1}{\sqrt{2m}} \overline{Q}^{\dot{1}} |0, 0\rangle = |\tfrac{1}{2}, \tfrac{1}{2}\rangle,$$

$$\frac{1}{\sqrt{2m}} \overline{Q}^{\dot{2}} |0, 0\rangle = |\tfrac{1}{2}, -\tfrac{1}{2}\rangle,$$

$$\frac{1}{\sqrt{2m}} \overline{Q}^{\dot{1}} |\tfrac{1}{2}, \tfrac{1}{2}\rangle = |1, 1\rangle,$$

$$\frac{1}{\sqrt{2m}} \overline{Q}^{\dot{2}} |\tfrac{1}{2}, -\tfrac{1}{2}\rangle = |1, -1\rangle,$$

$$\frac{1}{\sqrt{2m}} \overline{Q}^{\dot{2}} |\tfrac{1}{2}, \tfrac{1}{2}\rangle = \frac{1}{\sqrt{2}} [|1, 0\rangle + |0, 0\rangle],$$

$$\frac{1}{\sqrt{2m}} \overline{Q}^{\dot{1}} |\tfrac{1}{2}, -\tfrac{1}{2}\rangle = \frac{1}{\sqrt{2}} [|1, 0\rangle - |0, 0\rangle].$$

Solution

In the rest frame

$$P^\mu = (m, 0), \ W^\mu = \frac{1}{2}\epsilon^{\mu 0 \rho \gamma} P_0 M_{\rho \gamma},$$

$$W^0 = 0, \ W^i = \frac{m}{2}\epsilon^{ijk} M_{jk} = mJ^i,$$

$$\left[J^i, Q^\alpha\right] = -\frac{1}{2}(\sigma^i)^\alpha{}_\beta Q^\beta,$$

$$\left[J^i, \bar{Q}_{\dot\alpha}\right] = -\frac{1}{2}(\sigma^i)_{\dot\alpha}{}^{\dot\beta}\bar{Q}_{\dot\beta}.$$

Thus

$$\left[J^3, Q^1\right] = -\frac{1}{2}Q^1, \left[J^+, Q^1\right] = -Q^2, \left[J^-, Q^1\right] = 0,$$

$$\left[J^3, Q^2\right] = \frac{1}{2}Q^2, \left[J^+, Q^2\right] = 0, \left[J^-, Q^2\right] = -Q^1.$$

Now

$$\bar{Q}_{\dot\alpha} = \epsilon_{\dot\alpha \dot\beta}\bar{Q}^{\dot\beta}, \bar{Q}_{\dot 1} = \bar{Q}^{\dot 2}, \ \bar{Q}_{\dot 2} = -\bar{Q}^{\dot 1}.$$

Thus

$$\left[J^3, \bar{Q}_{\dot 1}\right] = -\frac{1}{2}\bar{Q}_{\dot 1} \rightarrow \left[J^3, \bar{Q}^{\dot 2}\right] = -\frac{1}{2}\bar{Q}^{\dot 2},$$

$$\left[J^+, \bar{Q}_{\dot 1}\right] = \bar{Q}_{\dot 2} \rightarrow \left[J^+, \bar{Q}^{\dot 2}\right] = -\frac{1}{2}\bar{Q}^{\dot 1},$$

$$\left[J^-, \bar{Q}_{\dot 1}\right] = 0 \rightarrow \left[J^-, \bar{Q}^{\dot 2}\right], = 0$$

$$\left[J^3, \bar{Q}_{\dot 2}\right] = \frac{1}{2}\bar{Q}_{\dot 2} \rightarrow \left[J^3, \bar{Q}^{\dot 1}\right] = \frac{1}{2}\bar{Q}^{\dot 1},$$

$$\left[J^+, \bar{Q}_{\dot 2}\right] = 0 \rightarrow \left[J^+, \bar{Q}^{\dot 1}\right] = 0,$$

$$\left[J^-, \bar{Q}_{\dot 2}\right] = -\bar{Q}_{\dot 1} \rightarrow \left[J^-, \bar{Q}^{\dot 1}\right] = \bar{Q}^{\dot 2},$$

$$J^2 = J^+ J^- - J^3 + \left(J^3\right)^2 = J^- J^+ + J^3 + \left(J^3\right)^2.$$

Now

$$J^3 \bar{Q}^{\dot 2}|j, j\rangle = \left(j\bar{Q}^{\dot 2} - \frac{1}{2}\bar{Q}^{\dot 2}\right)|j, j\rangle$$

$$= \left(j - \frac{1}{2}\right)\bar{Q}^{\dot 2}|j, j\rangle.$$

Similarly

$$J^3 \bar{Q}^{\dot{2}}|j, -j\rangle = \left(-j - \frac{1}{2}\right) \bar{Q}^{\dot{2}}|j, -j\rangle,$$

$$J^3 \bar{Q}^{\dot{1}}|j, j\rangle = \left(j + \frac{1}{2}\right) \bar{Q}^{\dot{1}}|j, j\rangle,$$

$$J^3 \bar{Q}^{\dot{1}}|j, -j\rangle = \left(-j + \frac{1}{2}\right) \bar{Q}^{\dot{1}}|j, j\rangle.$$

Also

$$\left[J^-, \bar{Q}^{\dot{2}}\right] = 0 \to J^- \bar{Q}^{\dot{2}}|j, -j\rangle = 0,$$

$$\left[J^+, \bar{Q}^{\dot{1}}\right] = 0 \to J^+ \bar{Q}^{\dot{1}}|j, j\rangle = 0.$$

So

$$J^2 \bar{Q}^{\dot{2}}|j, -j\rangle = \left(J^+ J^- - J^3 + \left(J^3\right)^2\right) \bar{Q}^{\dot{2}}|j, -j\rangle$$

$$= \left(j + \frac{1}{2}\right)\left(j + \frac{3}{2}\right) \bar{Q}^{\dot{2}}|j, -j\rangle,$$

$$J^2 \bar{Q}^{\dot{1}}|j, j\rangle = \left(J^- J^+ + J^3 + \left(J^3\right)^2\right) \bar{Q}^{\dot{1}}|j, j\rangle$$

$$= \left(j + \frac{1}{2}\right)\left(j + \frac{3}{2}\right) \bar{Q}^{\dot{1}}|j, j\rangle.$$

Thus

$$\frac{1}{\sqrt{2m}} \bar{Q}^{\dot{2}}|j, -j\rangle = |j + \frac{1}{2}, -\left(j + \frac{1}{2}\right)\rangle,$$

$$\frac{1}{\sqrt{2m}} \bar{Q}^{\dot{1}}|j, j\rangle = |j + \frac{1}{2}, j + \frac{1}{2}\rangle.$$

Hence

$$\frac{1}{\sqrt{2m}} \bar{Q}^{\dot{1}}|0, 0\rangle = |\frac{1}{2}, \frac{1}{2}\rangle,$$

$$\frac{1}{\sqrt{2m}} \bar{Q}^{\dot{2}}|0, 0\rangle = |\frac{1}{2}, -\frac{1}{2}\rangle,$$

$$\frac{1}{\sqrt{2m}} \bar{Q}^{\dot{1}}|\frac{1}{2}, \frac{1}{2}\rangle = |1, 1\rangle,$$

$$\frac{1}{\sqrt{2m}} \bar{Q}^{\dot{2}}|\frac{1}{2}, -\frac{1}{2}\rangle = |1, -1\rangle,$$

$$\frac{1}{\sqrt{2m}}\overline{Q}^{\dot{2}}|\frac{1}{2},\frac{1}{2}\rangle = |1,0\rangle,$$

$$\frac{1}{\sqrt{2m}}\overline{Q}^{\dot{1}}|\frac{1}{2},-\frac{1}{2}\rangle = |1,0\rangle,$$

$$\frac{1}{\sqrt{2m}}\overline{Q}^{\dot{2}}|\frac{1}{2},\frac{1}{2}\rangle = \frac{1}{\sqrt{2}}[|1,0\rangle + |0,0\rangle],$$

$$\frac{1}{\sqrt{2m}}\overline{Q}^{\dot{1}}|\frac{1}{2},-\frac{1}{2}\rangle = \frac{1}{\sqrt{2}}[|1,0\rangle - |0,0\rangle].$$

8.17 Q_α and $\overline{Q}_{\dot\alpha}$ are generators of SUSY transformation. For a chiral supermultiplet (ξ,ϕ), the supersymmetric transformation is given by

$$\delta_\epsilon \phi = \sqrt{2}\epsilon^T C\xi, \quad \delta_\epsilon \xi = \sqrt{2}i\sigma^\mu C\partial_\mu \phi \epsilon^*$$

where ϵ is a parameter which transforms as a left-handed chiral spinor.

i) Show that

$$[\delta_{\epsilon_1},\delta_{\epsilon_2}]\phi = 2i[\epsilon_1^\alpha (\sigma^\mu)_{\alpha\dot\beta}\bar\epsilon_2^{\dot\beta} - \epsilon_2^\alpha (\sigma^\mu)_{\alpha\dot\beta}\bar\epsilon_1^{\dot\beta}]\partial_\mu \phi.$$

ii) The SUSY transformation in terms of supersymmetric charges Q and $\overline{Q}$ can be written as

$$\delta_\epsilon = \epsilon \cdot Q + \bar\epsilon \cdot \overline{Q} = \epsilon_\alpha Q^\alpha + \bar\epsilon^{\dot\beta}\overline{Q}_{\dot\beta}$$

$$= -\epsilon^\alpha Q_\alpha + \bar\epsilon^{\dot\beta}\overline{Q}_{\dot\beta}.$$

Show that

$$[\delta_{\epsilon_1},\delta_{\epsilon_2}] = \epsilon_1^\alpha \{Q_\alpha,\overline{Q}_{\dot\beta}\}\bar\epsilon_2^{\dot\beta} - \epsilon_2^\alpha \{Q_\alpha,\overline{Q}_{\dot\beta}\}\bar\epsilon_1^{\dot\beta}.$$

Hence show that we get

$$[\delta_{\epsilon_1},\delta_{\epsilon_2}]\phi = 2i[\epsilon_1^\alpha (\sigma^\mu)_{\alpha\dot\beta}\bar\epsilon_2^{\dot\beta} - \epsilon_2^\alpha (\sigma^\mu)_{\alpha\dot\beta}\bar\epsilon_1^{\dot\beta}]\partial_\mu \phi.$$

Solution

(i)

$$\delta_{\epsilon_1}\delta_{\epsilon_2}\phi = \sqrt{2}\epsilon_2^T C\delta_{\epsilon_1}\xi = 2\epsilon_2^T C\sigma^\mu C\epsilon_1^* \partial_\mu \phi,$$

$$C\sigma^\mu C = -\sigma^2 \sigma^\mu \sigma^2 = -\bar\sigma^\mu.$$

Thus

$$\delta_{\epsilon_1}\delta_{\epsilon_2}\phi = -2i\epsilon_2^T \bar\sigma^{\mu T}\epsilon_1^* \partial_\mu \phi.$$

Therefore

$$[\delta_{\epsilon_1}, \delta_{\epsilon_2}]\phi = -2i\left[\epsilon_2^T\bar{\sigma}^{\mu T}\epsilon_1^* - \epsilon_1^T\bar{\sigma}^{\mu T}\epsilon_2^*\right]\partial_\mu\phi$$

$$= 2i\left[\left(\epsilon_1^\dagger\bar{\sigma}^{\mu T}\epsilon_2\right)^T - \left(\epsilon_2^\dagger\bar{\sigma}^{\mu T}\epsilon_1\right)^T\right]\partial_\mu\phi$$

$$2i\left[\epsilon_1^\dagger\bar{\sigma}^{\mu T}\epsilon_2 - \epsilon_2^\dagger\bar{\sigma}^{\mu T}\epsilon_1\right]\partial_\mu\phi$$

$$= 2i\left[(\epsilon_1^*)^T\bar{\sigma}^{\mu T}\epsilon_2 - (1\longleftrightarrow 2)\right]\partial_\mu\phi$$

$$= 2i\left[\bar{\epsilon}_1^{\dot{\beta}}\left(\bar{\sigma}^\mu\right)_{\dot{\beta}\alpha}\epsilon_2^\alpha - (1\longleftrightarrow 2)\right]\partial_\mu\phi$$

$$= 2i[\epsilon_1^\alpha(\sigma^\mu)_{\alpha\dot{\beta}}\bar{\epsilon}_2^{\dot{\beta}} - \epsilon_2^\alpha(\sigma^\mu)_{\alpha\dot{\beta}}\bar{\epsilon}_1^{\dot{\beta}}]\partial_\mu\phi.$$

(ii)

$$\delta_\epsilon = \epsilon\cdot Q + \bar{\epsilon}\cdot\overline{Q} = \epsilon_\alpha Q^\alpha + \bar{\epsilon}^{\dot{\beta}}\overline{Q}_{\dot{\beta}},$$

$$\epsilon_\alpha Q^\alpha = \varepsilon_{\alpha\gamma}\epsilon^\gamma\varepsilon^{\alpha\beta}Q_\beta = -\delta_\gamma^\beta\epsilon^\gamma Q_\beta = -\epsilon^\gamma Q_\gamma = -\epsilon^\alpha Q_\alpha.$$

Thus

$$\delta_{\epsilon_1}\delta_{\epsilon_2} - \delta_{\epsilon_2}\delta_{\epsilon_1} = -\epsilon_1^\alpha\left\{Q_\alpha, Q_\alpha\right\}\epsilon_2^\alpha - \epsilon_1^{\dot{\beta}}\left\{\overline{Q}_{\dot{\beta}}, \overline{Q}_{\dot{\beta}}\right\}\bar{\epsilon}_2^{\dot{\beta}}$$

$$- \epsilon_1^\alpha\left\{Q_\alpha, \overline{Q}_{\dot{\beta}}\right\}\bar{\epsilon}_2^{\dot{\beta}} + \epsilon_2^\alpha\left\{Q_\alpha, \overline{Q}_{\dot{\beta}}\right\}\bar{\epsilon}_1^{\dot{\beta}}$$

$$= -\left[\epsilon_1^\alpha\left\{Q_\alpha, \overline{Q}_{\dot{\beta}}\right\}\bar{\epsilon}_2^{\dot{\beta}} - \epsilon_2^\alpha\left\{Q_\alpha, \overline{Q}_{\dot{\beta}}\right\}\bar{\epsilon}_1^{\dot{\beta}}\right].$$

Therefore

$$[\delta_{\epsilon_1}, \delta_{\epsilon_2}]\phi = -2\left[\epsilon_1^\alpha(\sigma^\mu)_{\alpha\dot{\beta}}\bar{\epsilon}_2^{\dot{\beta}} - \epsilon_2^\alpha(\sigma^\mu)_{\alpha\dot{\beta}}\bar{\epsilon}_1^{\dot{\beta}}\right]P_\mu$$

and

$$[\delta_{\epsilon_1}, \delta_{\epsilon_2}]\phi = 2i\left[\epsilon_1^\alpha(\sigma^\mu)_{\alpha\dot{\beta}}\bar{\epsilon}_2^{\dot{\beta}} - \epsilon_2^\alpha(\sigma^\mu)_{\alpha\dot{\beta}}\bar{\epsilon}_1^{\dot{\beta}}\right]\partial_\mu\phi.$$

8.18 For a chiral supermultiplet, the Lagrangian is given by

$$L = \frac{1}{2}\partial^\mu\phi\partial_\mu\phi^* + i\xi^\dagger\bar{\sigma}^\mu\partial_\mu\xi. \qquad (I)$$

Show that the Lagrangian is invariant under the infinitesimal supersymmetric transformations

$$\delta_\epsilon\phi = \sqrt{2}\epsilon^T C\xi, \quad \delta_\epsilon\xi = \sqrt{2}i\sigma^\mu C\partial_\mu\phi\epsilon^*.$$

PART 2

General Theory of Relativity: Riemannian Geometry; Curved Space Time

Chapter 9

TENSOR ANALYSIS AND AFFINE CONNECTION

9.1 Introduction

Matter is the source of gravity; this was expressed by Newton as law of gravitational attraction:

$$\mathbf{F} = -G\frac{mM}{r^2}\mathbf{e}_r, \tag{9.1}$$

i.e., the force between two objects of masses m and M is given by Eq. (9.1). Here G is the Newton's gravitational constant. On the other hand, the force between two charged particles can be expressed in terms of fields. A charged particle produces an electric field $\mathbf{E}$ and also a magnetic field $\mathbf{B}$ when moving. In the presence of electromagnetic fields $\mathbf{E}$ and $\mathbf{B}$, a particle of charge e experiences a force

$$\mathbf{F} = e[\mathbf{E} + \frac{1}{c}\mathbf{v} \times \mathbf{B}], \tag{9.2}$$

where c is the speed of light in the vacuum.

Is it possible to replace Newton's law of gravitational attraction by a law expressed in terms of gravitational field? Einstein showed that it is possible in the form of his general theory of relativity.

Basic laws governing electromagnetic fields are Maxwell's equations which can be expressed in covariant form as

$$\partial_\mu F^{\mu\nu} = J^\nu. \tag{9.3}$$

The law expressed in the form of the above equations is invariant under Lorentz transformations connecting inertial frames. Such a transformation leaves the distance ds of two infinitesimally adjacent points ξ^α and $\xi^\alpha + d\xi^\alpha$:

$$ds^2 = \eta^{\alpha\beta}d\xi_\alpha d\xi_\beta = \eta_{\alpha\beta}d\xi^\alpha d\xi^\beta \tag{9.4}$$

invariant. Here $\eta^{\alpha\beta}$ are constants with components $(1, -1, -1, -1)$ and ξ^α are space-time coordinates in inertial coordinate system. Also

$$d\xi'^\alpha = \frac{\partial \xi'^\alpha}{\partial \xi^\beta} d\xi^\beta.$$

The invariance of physical laws under Lorentz transformations connecting inertial frames was extended by Einstein by demanding that all physical laws to be invariant under general coordinate transformations defined by

$$x'^\mu = x'^\mu(x^0, x^i)$$

$$dx'^\mu = \sum_\nu \frac{\partial x'^\mu}{\partial x^\nu} dx^\nu \tag{9.5}$$

which leaves the quadratic form

$$ds^2 = g^{\mu\nu} dx_\mu dx_\nu = g_{\mu\nu} dx^\mu dx^\nu \tag{9.6}$$

invariant. The metric $g^{\mu\nu}$ which are continuous functions of x^μ determine the metric in four-dimensional manifold. But they are not given a priori. Einstein introduced a novel idea that it is the gravitational field which determines the metric field $g_{\mu\nu}$. Existence of matter caused the fabric of space-time to wrap somewhat like the effect of bouncing ball placed on foam. Gravity resides in the curvature of space-time [$g_{\mu\nu}$ introduced above]. The geometry which describes curved spaces is known as Riemannian geometry. It is the physics that determines the geometry. To conclude: The quadratic differential form $g_{\mu\nu} dx^\mu dx^\nu$ with the metric tensor $g_{\mu\nu}(x)$ identified with gravitational potential is the distinguishing feature of general theory of relativity in contrast with electrodynamics where we have linear differential form $A_\mu(x) dx^\mu$.

9.2   Metric Tensor, Tensors, Tensor Densities

In the inertial coordinate system, we denote the 4-vector

$$\xi^\alpha = \left(\xi^0, \xi^i \right), \, \xi_\alpha = (\xi_0, \xi_i)$$

with a metric tensor

$$\eta_{\alpha\beta} = diag\,(1, -1, -1, -1) = \eta^{\alpha\beta}$$

$$\det \eta_{\alpha\beta} = -1.$$

The distance ds between two infinitesimal adjacent points ξ^α, $\xi^\alpha + d\xi^\alpha$ in inertial coordinate system is given by

$$ds^2 = \eta^{\alpha\beta} d\xi_\alpha d\xi_\beta = \eta_{\alpha\beta} d\xi^\alpha d\xi^\beta. \tag{9.7}$$

We denote the general coordinate system by a 4-vector

$$x^\mu = \left(x^0, x^i\right), \ x_\mu = (x_0, x_i).$$

The transformation from ξ to x is given by:

$$dx^\mu = \frac{\partial x^\mu}{\partial \xi^\alpha} d\xi^\alpha,$$

$$dx_\mu = \frac{\partial \xi^\alpha}{\partial x^\mu} d\xi_\alpha. \tag{9.8}$$

Note that the transformation is between the differentials. It is possible to introduce inertial frame locally in the curved space.

The Jacobian of transformation is

$$J = \left| \frac{\partial x}{\partial \xi} \right| = \varepsilon_{\mu\nu\rho\lambda} \frac{\partial x^\mu}{\partial \xi^0} \frac{\partial x^\nu}{\partial \xi^1} \frac{\partial x^\rho}{\partial \xi^2} \frac{\partial x^\lambda}{\partial \xi^3}. \tag{9.9}$$

The inverse transformation is

$$d\xi^\alpha = \frac{\partial \xi^\alpha}{\partial x^\mu} dx^\mu, \ d\xi_\alpha = \frac{\partial x^\mu}{\partial \xi^\alpha} dx_\mu \tag{9.10}$$

and the corresponding Jacobian is given by

$$J^{-1} = \left| \frac{\partial \xi}{\partial x} \right| = \varepsilon_{\alpha\beta\gamma\delta} \frac{\partial \xi^\alpha}{\partial x^0} \frac{\partial \xi^\beta}{\partial x^1} \frac{\partial \xi^\gamma}{\partial x^2} \frac{\partial \xi^\delta}{\partial x^3}. \tag{9.11}$$

The metric tensor $g_{\mu\nu}$ is characteristic of curvature.

From Eq. (9.7), using Eq. (9.10):

$$ds^2 = \eta_{\alpha\beta} \frac{\partial \xi^\alpha}{\partial x^\mu} dx^\mu \frac{\partial \xi^\beta}{\partial x^\nu} dx^\nu$$

$$= \left(\eta_{\alpha\beta} \frac{\partial \xi^\alpha}{\partial x^\mu} \frac{\partial \xi^\beta}{\partial x^\nu} \right) dx^\mu dx^\nu$$

$$= g_{\mu\nu} dx^\mu dx^\nu, \tag{9.12}$$

where

$$g_{\mu\nu} = \eta_{\alpha\beta} \frac{\partial \xi^\alpha}{\partial x^\mu} \frac{\partial \xi^\beta}{\partial x^\nu} \tag{9.13}$$

is the metric tensor. It is a function of x.

Now

$$g \equiv \det\left(g_{\mu\nu}\right) = \det\left(\eta_{\alpha\beta} \frac{\partial \xi^\alpha}{\partial x^\mu} \frac{\partial \xi^\beta}{\partial x^\nu} \right)$$

$$= -J^{-2} \tag{9.14}$$

or

$$J^{-1} = \sqrt{-g}, \ J = \frac{1}{\sqrt{-g}}. \tag{9.15}$$

Alternatively

$$ds^2 = \eta^{\alpha\beta} d\xi_\alpha d\xi_\beta$$
$$= \left(\eta^{\alpha\beta} \frac{\partial x^\mu}{\partial \xi^\alpha} \frac{\partial x^\nu}{\partial \xi^\beta} \right) dx_\mu dx_\nu$$
$$= g^{\mu\nu} dx_\mu dx_\nu, \tag{9.16}$$

where

$$g^{\mu\nu} = \eta^{\alpha\beta} \frac{\partial x^\mu}{\partial \xi^\alpha} \frac{\partial x^\nu}{\partial \xi^\beta}. \tag{9.17}$$

Thus

$$g^{\mu\lambda} g_{\lambda\nu} = \delta^\mu_\nu, \tag{9.18}$$

where δ^μ_ν is Krönecker delta function.

9.2.1 *Tensors, Tensor Densities*

In curved space we cannot expect linear transformations between coordinates themselves. Only differentials dx'^μ and dx^μ are connected by general coordinate transformation viz

$$dx'^\mu = \frac{\partial x'^\mu}{\partial x^\lambda} dx^\lambda. \tag{9.19}$$

Inverse transformation:

$$dx^\mu = \frac{\partial x^\mu}{\partial x'^\lambda} dx'^\lambda. \tag{9.20}$$

The Jacobian of the transformation x to x' is $\left| \frac{\partial x'}{\partial x} \right|$.

Any quantity which has the transformation law

$$V'^\mu = \frac{\partial x'^\mu}{\partial x^\lambda} V^\lambda, \tag{9.21}$$

where $\frac{\partial x'^\mu}{\partial x^\lambda}$ must be evaluated at a point $P(x^\mu)$ is called a contravariant vector. A covariant vector V_μ transforms as

$$V'_\mu = \frac{\partial x^\lambda}{\partial x'^\mu} V_\lambda. \tag{9.22}$$

For a scalar function

$$f'(x'^\mu) = f(x^\mu)$$

thus

$$\frac{\partial f}{\partial x'^\mu} = \frac{\partial f}{\partial x^\lambda} \frac{\partial x^\lambda}{\partial x'^\mu} = \frac{\partial x^\lambda}{\partial x'^\mu} \frac{\partial f}{\partial x^\lambda}, \tag{9.23}$$

i.e., transforms as covariant vector.

Contravariant tensor $T^{\mu\nu}$ and covariant tensor $T_{\mu\nu}$ are tensors of rank 2, each with 16 components and they transform as

$$T'^{\mu\nu} = \frac{\partial x'^{\mu}}{\partial x^{\lambda}} \frac{\partial x'^{\nu}}{\partial x^{\rho}} T^{\lambda\rho} , \qquad (9.24)$$

$$T'_{\mu\nu} = \frac{\partial x^{\lambda}}{\partial x'^{\mu}} \frac{\partial x^{\rho}}{\partial x'^{\nu}} T_{\lambda\rho} . \qquad (9.25)$$

A mixed tensor of rank 2, T^{μ}_{ν} transforms as

$$T'^{\mu}_{\nu} = \frac{\partial x'^{\mu}}{\partial x^{\lambda}} \frac{\partial x^{\rho}}{\partial x'^{\nu}} T^{\lambda}_{\rho}. \qquad (9.26)$$

In particular (cf. Eq. (9.23))

$$\begin{aligned}
g'_{\mu\nu} &= \eta_{\alpha\beta} \frac{\partial x^{\lambda}}{\partial x'^{\mu}} \frac{\partial \xi^{\alpha}}{\partial x^{\lambda}} \frac{\partial x^{\sigma}}{\partial x'^{\nu}} \frac{\partial \xi^{\beta}}{\partial x^{\sigma}} \\
&= \frac{\partial x^{\lambda}}{\partial x'^{\mu}} \frac{\partial x^{\sigma}}{\partial x'^{\nu}} \left(\eta_{\alpha\beta} \frac{\partial \xi^{\alpha}}{\partial x^{\lambda}} \frac{\partial \xi^{\beta}}{\partial x^{\sigma}} \right) \\
&= \frac{\partial x^{\lambda}}{\partial x'^{\mu}} \frac{\partial x^{\sigma}}{\partial x'^{\nu}} g_{\lambda\sigma}.
\end{aligned} \qquad (9.27)$$

Similarly

$$g'^{\mu\nu} = \frac{\partial x'^{\mu}}{\partial x^{\lambda}} \frac{\partial x'^{\nu}}{\partial x^{\sigma}} g^{\lambda\sigma}. \qquad (9.28)$$

Krönecker δ^{μ}_{ν} transforms as mixed tensor

$$\begin{aligned}
\delta'^{\mu}_{\nu} &= \frac{\partial x'^{\mu}}{\partial x^{\lambda}} \frac{\partial x^{\rho}}{\partial x'^{\nu}} \delta^{\lambda}_{\rho} \\
&= \frac{\partial x'^{\mu}}{\partial x^{\lambda}} \frac{\partial x^{\lambda}}{\partial x'^{\nu}} = \frac{\partial x'^{\mu}}{\partial x'^{\nu}} = \delta'^{\mu}_{\nu}.
\end{aligned} \qquad (9.29)$$

Now

$$\begin{aligned}
g'_{\mu\nu} &= \frac{\partial x^{\lambda}}{\partial x'^{\mu}} \frac{\partial x^{\sigma}}{\partial x'^{\nu}} g_{\lambda\sigma} \\
&= \frac{\partial x^{\lambda}}{\partial x'^{\mu}} g_{\lambda\sigma} \frac{\partial x^{\sigma}}{\partial x'^{\nu}}.
\end{aligned}$$

Taking the determinant on both sides

$$g' = \left| \frac{\partial x}{\partial x'} \right|^{2} g = \left| \frac{\partial x'}{\partial x} \right|^{-2} g \qquad (9.30)$$

g' is called a scalar density of weight -2.

Now

$$\frac{\partial x^{\rho}}{\partial x'^{\mu}} \frac{\partial x^{\varkappa}}{\partial x'^{\nu}} \frac{\partial x^{\gamma}}{\partial x'^{\lambda}} \frac{\partial x^{\delta}}{\partial x'^{\sigma}} \varepsilon_{\rho\varkappa\gamma\delta} \propto \varepsilon'_{\mu\nu\lambda\sigma} = a\varepsilon'_{\mu\nu\lambda\sigma}. \qquad (9.31)$$

To determine a, take $\mu = 0$, $\nu = 1$, $\lambda = 2$, $\sigma = 3$, thus

$$a = \varepsilon_{\rho\varkappa\gamma\delta} \frac{\partial x^\rho}{\partial x'^0} \frac{\partial x^\varkappa}{\partial x'^1} \frac{\partial x^\gamma}{\partial x'^2} \frac{\partial x^\delta}{\partial x'^3} = \left| \frac{\partial x}{\partial x'} \right| = \left| \frac{\partial x'}{\partial x} \right|^{-1}. \tag{9.32}$$

Hence

$$\varepsilon'_{\mu\nu\lambda\sigma} = \left| \frac{\partial x'}{\partial x} \right| \frac{\partial x^\rho}{\partial x'^\mu} \frac{\partial x^\varkappa}{\partial x'^\nu} \frac{\partial x^\gamma}{\partial x'^\lambda} \frac{\partial x^\delta}{\partial x'^\sigma} \varepsilon_{\rho\varkappa\gamma\delta} \tag{9.33}$$

is a tensor density of weight 1 and

$$\varepsilon'^{\mu\nu\lambda\sigma} = \left| \frac{\partial x'}{\partial x} \right|^{-1} \frac{\partial x'^\mu}{\partial x^\rho} \frac{\partial x'^\nu}{\partial x^\varkappa} \frac{\partial x'^\lambda}{\partial x^\gamma} \frac{\partial x'^\sigma}{\partial x^\delta} \varepsilon^{\rho\varkappa\gamma\delta} \tag{9.34}$$

is a tensor density of weight -1.

Now

$$\left| \frac{\partial x'}{\partial x} \right| = \left| \frac{\partial x'/\partial \xi}{\partial x/\partial \xi} \right| = \frac{J'}{J} = \frac{\sqrt{-g}}{\sqrt{-g'}}. \tag{9.35}$$

Thus from Eqs. (9.33) and (9.34)

$$\sqrt{-g'}\varepsilon'_{\mu\nu\lambda\sigma} = \sqrt{-g} \frac{\partial x^\rho}{\partial x'^\mu} \frac{\partial x^\varkappa}{\partial x'^\nu} \frac{\partial x^\gamma}{\partial x'^\lambda} \frac{\partial x^\delta}{\partial x'^\sigma} \varepsilon_{\rho\varkappa\gamma\delta}. \tag{9.36}$$

Hence

$$\tilde{\varepsilon}_{\mu\nu\lambda\sigma} = \sqrt{-g}\varepsilon_{\mu\nu\lambda\sigma} \tag{9.37}$$

and

$$\tilde{\varepsilon}^{\mu\nu\lambda\sigma} = \frac{1}{\sqrt{-g}}\varepsilon^{\mu\nu\lambda\sigma} \tag{9.38}$$

are covariant and contravariant tensors of rank 4.

How does the volume element

$$d^4x = dx^0 dx^1 dx^2 dx^3$$

transforms?

From calculus

$$d^4x' = \left| \frac{\partial x'}{\partial x} \right| d^4x$$

$$= \frac{\sqrt{-g}}{\sqrt{-g'}} d^4x. \tag{9.39}$$

Hence

$$\sqrt{-g'} d^4x' = \sqrt{-g} d^4x \tag{9.40}$$

is invariant.

Contraction: Tensors can be contracted by setting two indices equal provided one index is upstairs and other is downstairs, e.g.,

$$U^\mu = T^{\mu\nu} V_\nu$$

$$U'^\mu = \frac{\partial x'^\mu}{\partial x^\lambda} \frac{\partial x'^\nu}{\partial x^\rho} T^{\lambda\rho} \frac{\partial x^\sigma}{\partial x'^\nu} V_\sigma$$

$$= \frac{\partial x'^\mu}{\partial x^\lambda} \frac{\partial x'^\nu}{\partial x^\rho} \frac{\partial x^\sigma}{\partial x'^\nu} T^{\lambda\rho} V_\sigma$$

$$= \frac{\partial x'^\mu}{\partial x^\lambda} \delta^\sigma_\rho T^{\lambda\rho} V_\sigma$$

$$= \frac{\partial x'^\mu}{\partial x^\lambda} T^{\lambda\sigma} V_\sigma = \frac{\partial x'^\mu}{\partial x^\lambda} U^\lambda. \tag{9.41}$$

9.3 Covariant Derivative, Affine Connection, Christoffel Symbol

The derivative of a vector in inertial frame transforms as mixed tensor, but this is not so in general coordinate system due to curvature.

Now

$$\frac{\partial V'^\lambda}{\partial x'^\mu} = \frac{\partial x^\sigma}{\partial x'^\mu} \frac{\partial}{\partial x^\sigma} \left(\frac{\partial x'^\lambda}{\partial x^\rho} V^\rho \right)$$

$$= \frac{\partial x^\sigma}{\partial x'^\mu} \frac{\partial x'^\lambda}{\partial x^\rho} \frac{\partial V^\rho}{\partial x^\sigma} + \frac{\partial x^\sigma}{\partial x'^\mu} V^\rho \frac{\partial^2 x'^\lambda}{\partial x^\sigma \partial x^\rho}. \tag{9.42}$$

Due to second term in Eq. (9.42), $\frac{\partial V^\lambda}{\partial x^\mu}$ does not transform as tensor. Similarly (see Problem 9.2):

$$\frac{\partial^2 f}{\partial x'^\mu \partial x'^\nu} = \frac{\partial x^\lambda}{\partial x'^\mu} \frac{\partial x^\sigma}{\partial x'^\nu} \frac{\partial^2 f}{\partial x^\lambda \partial x^\sigma} + \frac{\partial^2 x^\sigma}{\partial x'^\mu \partial x'^\nu} \frac{\partial f}{\partial x^\sigma}. \tag{9.43}$$

One defines now the derivative, called the covariant derivative ∇_μ to replace ∂_μ, so that $\nabla_\mu V^\nu$ transforms as a tensor. Let us write $\nabla_\mu V^\lambda$ in the following way

$$\nabla_\mu V^\lambda = V^\lambda_{;\mu} = \frac{\partial V^\lambda}{\partial x^\mu} + \Gamma^\lambda_{\mu\nu} V^\nu \tag{9.44}$$

where $\Gamma^\lambda_{\mu\nu}$ is called "affine connection" or Christoffel symbol. Christoffel symbol $\Gamma^\lambda_{\mu\nu}$ does not transform as a tensor and has the transformation property so that $V^\lambda_{;\mu}$ transforms as tensor.

Using expression for the metric tensor $g_{\mu\nu}$ given in Eq. (9.13) as guide, one may write

$$\Gamma^\lambda_{\mu\nu} = g^{\lambda\sigma} \eta_{\alpha\beta} \frac{\partial \xi^\alpha}{\partial x^\sigma} \frac{\partial^2 \xi^\beta}{\partial x^\mu \partial x^\nu}. \tag{9.45}$$

$\Gamma^\lambda_{\mu\nu}$ can be expressed in terms of derivative of metric tensor as follows:

$$g_{\sigma\mu} = \eta_{\alpha\beta} \frac{\partial \xi^\alpha}{\partial x^\sigma} \frac{\partial \xi^\beta}{\partial x^\mu},$$

$$\frac{\partial g_{\sigma\mu}}{\partial x^\nu} = \eta_{\alpha\beta} \frac{\partial^2 \xi^\alpha}{\partial x^\nu \partial x^\sigma} \frac{\partial \xi^\beta}{\partial x^\mu} + \eta_{\alpha\beta} \frac{\partial \xi^\alpha}{\partial x^\sigma} \frac{\partial^2 \xi^\beta}{\partial x^\mu \partial x^\nu},$$

$$\frac{\partial g_{\sigma\nu}}{\partial x^\mu} = \eta_{\alpha\beta} \frac{\partial^2 \xi^\alpha}{\partial x^\mu \partial x^\sigma} \frac{\partial \xi^\beta}{\partial x^\nu} + \eta_{\alpha\beta} \frac{\partial \xi^\alpha}{\partial x^\sigma} \frac{\partial^2 \xi^\beta}{\partial x^\mu \partial x^\nu},$$

$$\frac{\partial g_{\mu\nu}}{\partial x^\sigma} = \eta_{\alpha\beta} \frac{\partial^2 \xi^\alpha}{\partial x^\sigma \partial x^\mu} \frac{\partial \xi^\beta}{\partial x^\nu} + \eta_{\alpha\beta} \frac{\partial \xi^\alpha}{\partial x^\mu} \frac{\partial^2 \xi^\beta}{\partial x^\sigma \partial x^\nu}. \tag{9.46}$$

From Eqs. (9.45) and (9.46), we get

$$\Gamma^\lambda_{\mu\nu} = g^{\lambda\sigma} \eta_{\alpha\beta} \frac{\partial \xi^\alpha}{\partial x^\sigma} \frac{\partial^2 \xi^\beta}{\partial x^\mu \partial x^\nu}$$

$$= \frac{1}{2} g^{\lambda\sigma} \left[\frac{\partial g_{\sigma\mu}}{\partial x^\nu} + \frac{\partial g_{\sigma\nu}}{\partial x^\mu} - \frac{\partial g_{\mu\nu}}{\partial x^\sigma} \right] = \Gamma^\lambda_{\nu\mu}. \tag{9.47}$$

Using

$$g^{\lambda\sigma} = \eta^{\gamma\varkappa} \frac{\partial x^\lambda}{\partial \xi^\gamma} \frac{\partial x^\sigma}{\partial \xi^\varkappa}$$

along with Eq. (9.45), one can express $\Gamma^\lambda_{\mu\nu}$:

$$\Gamma^\lambda_{\mu\nu} = \frac{\partial x^\lambda}{\partial \xi^\alpha} \frac{\partial^2 \xi^\alpha}{\partial x^\mu \partial x^\nu}. \tag{9.48}$$

Now

$$\Gamma'^\lambda_{\mu\nu} = \frac{\partial x'^\lambda}{\partial \xi^\alpha} \frac{\partial^2 \xi^\alpha}{\partial x'^\mu \partial x'^\nu}$$

$$= \frac{\partial x'^\lambda}{\partial x^\rho} \frac{\partial x^\rho}{\partial \xi^\alpha} \left[\frac{\partial x^\varkappa}{\partial x'^\mu} \frac{\partial x^\sigma}{\partial x'^\nu} \frac{\partial^2 \xi^\alpha}{\partial x^\varkappa \partial x^\sigma} + \frac{\partial^2 x^\sigma}{\partial x'^\mu \partial x'^\nu} \frac{\partial \xi^\alpha}{\partial x^\sigma} \right],$$

where for $\frac{\partial^2 \xi^\alpha}{\partial x'^\mu \partial x'^\nu}$ the transformation law of the form given in Eq. (9.43) is used. Thus

$$\Gamma'^\lambda_{\mu\nu} = \frac{\partial x'^\lambda}{\partial x^\rho} \frac{\partial x^\varkappa}{\partial x'^\mu} \frac{\partial x^\sigma}{\partial x'^\nu} \Gamma^\rho_{\varkappa\sigma} + \frac{\partial x'^\lambda}{\partial x^\rho} \delta^\rho_\sigma \frac{\partial^2 x^\sigma}{\partial x'^\mu \partial x'^\nu}$$

$$= \frac{\partial x'^\lambda}{\partial x^\rho} \frac{\partial x^\varkappa}{\partial x'^\mu} \frac{\partial x^\sigma}{\partial x'^\nu} \Gamma^\rho_{\varkappa\sigma} + \frac{\partial x'^\lambda}{\partial x^\rho} \frac{\partial^2 x^\rho}{\partial x'^\mu \partial x'^\nu}. \tag{9.49}$$

Due to the presence of second term in Eq. (9.49), $\Gamma^\lambda_{\mu\nu}$ does not transform as tensor.

Now using the relations

$$\frac{\partial x'^\lambda}{\partial x^\rho}\frac{\partial x^\rho}{\partial x'^\nu} = \delta^\lambda_\nu$$

$$\frac{\partial}{\partial x'^\mu}\left(\frac{\partial x'^\lambda}{\partial x^\rho}\frac{\partial x^\rho}{\partial x'^\nu}\right) = 0,$$

we have

$$\frac{\partial x^\rho}{\partial x'^\nu}\frac{\partial^2 x'^\lambda}{\partial x'^\mu \partial x^\rho} = -\frac{\partial x'^\lambda}{\partial x^\rho}\frac{\partial^2 x^\rho}{\partial x'^\mu \partial x'^\nu}$$

or

$$\frac{\partial x'^\lambda}{\partial x^\rho}\frac{\partial^2 x^\rho}{\partial x'^\mu \partial x'^\nu} = -\frac{\partial x^\rho}{\partial x'^\nu}\frac{\partial x^\sigma}{\partial x'^\mu}\frac{\partial}{\partial x^\sigma}\left(\frac{\partial x'^\lambda}{\partial x^\rho}\right).$$

Hence

$$\Gamma'^\lambda_{\mu\nu} = \frac{\partial x'^\lambda}{\partial x^\rho}\frac{\partial x^\varkappa}{\partial x'^\mu}\frac{\partial x^\sigma}{\partial x'^\nu}\Gamma^\rho_{\varkappa\sigma} - \frac{\partial x^\sigma}{\partial x'^\mu}\frac{\partial x^\rho}{\partial x'^\nu}\frac{\partial^2 x'^\lambda}{\partial x^\sigma \partial x^\rho}. \tag{9.50}$$

Now we show that covariant derivative transforms as tensor

$$V'^\lambda_{;\mu} = \frac{\partial V'^\lambda}{\partial \acute{x}^\mu} + \Gamma'^\lambda_{\mu\nu}V'^\nu, \tag{9.51}$$

$$\frac{\partial V'^\lambda}{\partial \acute{x}^\mu} + \Gamma'^\lambda_{\mu\nu}V'^\nu = \frac{\partial x^\sigma}{\partial \acute{x}^\mu}\frac{\partial \acute{x}^\lambda}{\partial x^\rho}\frac{\partial V^\rho}{\partial x^\sigma} + \frac{\partial x^\sigma}{\partial \acute{x}^\mu}V^\rho\frac{\partial^2 \acute{x}^\lambda}{\partial x^\sigma \partial x^\rho}$$

$$+[\frac{\partial \acute{x}^\lambda}{\partial x^\rho}\frac{\partial x^\varkappa}{\partial \acute{x}^\mu}\frac{\partial x^\sigma}{\partial \acute{x}^\nu}\Gamma^\rho_{\varkappa\sigma} - \frac{\partial x^\sigma}{\partial \acute{x}^\mu}\frac{\partial x^\rho}{\partial \acute{x}^\nu}\frac{\partial^2 \acute{x}^\lambda}{\partial x^\sigma \partial x^\rho}][\frac{\partial \acute{x}^\nu}{\partial x^\tau}V^\tau]$$

$$= \frac{\partial x^\sigma}{\partial \acute{x}^\mu}\frac{\partial \acute{x}^\lambda}{\partial x^\rho}\frac{\partial V^\rho}{\partial x^\sigma} + \frac{\partial \acute{x}^\lambda}{\partial x^\rho}\frac{\partial x^\varkappa}{\partial \acute{x}^\mu}\delta^\sigma_\tau \Gamma^\rho_{\varkappa\sigma}V^\tau$$

$$+ \frac{\partial x^\sigma}{\partial \acute{x}^\mu}\frac{\partial^2 \acute{x}^\lambda}{\partial x^\sigma \partial x^\rho}V^\rho - \frac{\partial x^\sigma}{\partial \acute{x}^\mu}\delta^\rho_\tau\frac{\partial^2 \acute{x}^\lambda}{\partial x^\sigma \partial x^\rho}V^\tau$$

$$= \frac{\partial \acute{x}^\lambda}{\partial x^\rho}\frac{\partial x^\sigma}{\partial \acute{x}^\mu}[\frac{\partial V^\rho}{\partial x^\sigma} + \Gamma^\rho_{\sigma\kappa}V^\kappa]$$

$$= \frac{\partial \acute{x}^\lambda}{\partial x^\rho}\frac{\partial x^\sigma}{\partial \acute{x}^\mu}V^\rho_{;\sigma}. \tag{9.52}$$

In order to find the form of covariant derivative of covariant vector V_λ, we consider a scalar $U^\lambda V_\lambda$ for which

$$\nabla_\mu(U^\lambda V_\lambda) = U^\lambda \nabla_\mu V_\lambda + (\nabla_\mu U^\lambda)V_\lambda,$$

since for a scalar ∇_μ is identical with ∂_μ, that is

$$\partial_\mu(U^\lambda V_\lambda) = U^\lambda \nabla_\mu V_\lambda + (\nabla_\mu U^\lambda)V_\lambda$$

or

$$U^\lambda \frac{\partial V_\lambda}{\partial x^\mu} + (\frac{\partial U^\lambda}{\partial x^\mu})V_\lambda$$

$$= U^\lambda V_{\lambda;\mu} + U^\lambda_{;\mu} V_\lambda$$

$$= U^\lambda V_{\lambda;\mu} + (\frac{\partial U^\lambda}{\partial x^\mu} + \Gamma^\lambda_{\mu\nu}U^\nu)V_\lambda.$$

Thus

$$U^\lambda V_{\lambda;\mu} = U^\lambda \frac{\partial V_\lambda}{\partial x^\mu} - \Gamma^\lambda_{\mu\nu}U^\nu V_\lambda$$

$$= U^\lambda \frac{\partial V_\lambda}{\partial x^\mu} - \Gamma^\nu_{\mu\lambda}V_\nu U^\lambda,$$

where we have interchanged λ and ν in the second term on the right-hand side of the above equation. Since U^λ is arbitrary, we have

$$V_{\lambda;\mu} = \frac{\partial V_\lambda}{\partial x^\mu} - \Gamma^\nu_{\mu\lambda}V_\nu. \tag{9.53}$$

For the tensor $T^{\mu\nu}, T_{\mu\nu}, T^\mu_\nu$, covariant derivative is obtained as follows: Consider

$$T^{\mu\nu} = V^\mu V^\nu,$$

$$\nabla_\lambda T^{\mu\nu} = V^\mu \nabla_\lambda V^\nu + (\nabla_\lambda V^\mu) V^\nu$$

$$= V^\mu(\frac{\partial V^\nu}{\partial x^\lambda} + \Gamma^\nu_{\lambda\rho}V^\rho) + (\frac{\partial V^\mu}{\partial x^\lambda} + \Gamma^\mu_{\lambda\rho}V^\rho)V^\nu$$

$$= \frac{\partial}{\partial x^\lambda}(V^\mu V^\nu) + \Gamma^\nu_{\lambda\rho}V^\mu V^\rho + \Gamma^\mu_{\lambda\rho}V^\rho V^\nu$$

$$= \frac{\partial}{\partial x^\lambda}T^{\mu\nu} + \Gamma^\nu_{\lambda\rho}T^{\mu\rho} + \Gamma^\mu_{\lambda\rho}T^{\rho\nu}.$$

Thus

$$\nabla_\lambda T^{\mu\nu} = T^{\mu\nu}_{;\lambda}$$

$$= \frac{\partial T^{\mu\nu}}{\partial x^\lambda} + \Gamma^\mu_{\lambda\rho}T^{\rho\nu} + \Gamma^\nu_{\lambda\rho}T^{\mu\rho}. \tag{9.54}$$

Similarly (see Problem 9.3)

$$\nabla_\lambda T_{\mu\nu} = T_{\mu\nu;\lambda}$$

$$= \frac{\partial T_{\mu\nu}}{\partial x^\lambda} - \Gamma^\rho_{\lambda\mu}T_{\rho\nu} - \Gamma^\rho_{\lambda\nu}T_{\mu\rho}, \tag{9.55}$$

$$\nabla_\lambda T_\nu^\mu = T_{\nu;\lambda}^\mu$$
$$= \frac{\partial T_\nu^\mu}{\partial x^\lambda} + \Gamma_{\lambda\rho}^\mu T_\nu^\rho - \Gamma_{\lambda\nu}^\rho T_\rho^\mu. \tag{9.56}$$

In particular

$$\nabla_\lambda g^{\mu\nu} = g_{;\lambda}^{\mu\nu}$$
$$= \frac{\partial g^{\mu\nu}}{\partial x^\lambda} + \Gamma_{\lambda\rho}^\mu g^{\rho\nu} + \Gamma_{\lambda\rho}^\nu g^{\rho\mu}, \tag{9.57}$$

$$\nabla_\lambda g_{\mu\nu} = g_{\mu\nu;\lambda},$$
$$= \frac{\partial g_{\mu\nu}}{\partial x^\lambda} - \Gamma_{\lambda\mu}^\rho g_{\rho\nu} - \Gamma_{\lambda\nu}^\rho g_{\rho\mu}. \tag{9.58}$$

We now show that covariant derivative of metric tensor is zero.
    Now

$$V^\mu V_\mu = g^{\mu\nu} V_\nu V_\mu, \tag{9.59}$$
$$\nabla_\lambda (V^\mu V_\mu) = (\nabla_\lambda g^{\mu\nu}) V_\nu V_\mu + g^{\mu\nu} \nabla_\lambda (V_\nu V_\mu),$$

$$g^{\mu\nu} \nabla_\lambda (V_\nu V_\mu) = g^{\mu\nu} [(\nabla_\lambda V_\nu) V_\mu + V_\nu \nabla_\lambda V_\mu],$$
$$= [(\nabla_\lambda V_\nu) V^\nu + V^\mu \nabla_\lambda V_\mu]$$
$$= [(\nabla_\lambda V_\mu) V^\mu + V^\mu \nabla_\lambda V_\mu]$$
$$= \nabla_\lambda (V^\mu V_\mu). \tag{9.60}$$

Hence from Eqs. (9.59) and (9.60)

$$g_{;\lambda}^{\mu\nu} = 0. \tag{9.61}$$

Similarly

$$g_{\mu\nu;\lambda} = 0. \tag{9.62}$$

For alternative derivative see Problem 9.4.
    The vanishing of

$$g_{\mu\nu;\lambda} = 0,$$

implies

$$0 = \partial_\lambda g_{\mu\nu} - \Gamma_{\lambda\mu}^\sigma g_{\sigma\nu} - \Gamma_{\lambda\nu}^\sigma g_{\sigma\mu}. \tag{9.63}$$

Let us cyclically permute indices

$$0 = \partial_\mu g_{\nu\lambda} - \Gamma_{\mu\nu}^\sigma g_{\sigma\lambda} - \Gamma_{\mu\lambda}^\sigma g_{\sigma\nu}, \tag{9.64}$$

$$0 = \partial_\nu g_{\lambda\mu} - \Gamma^\sigma_{\nu\lambda} g_{\sigma\mu} - \Gamma^\sigma_{\nu\mu} g_{\sigma\lambda}. \tag{9.65}$$

Add Eqs. (9.64) and (9.65) and subtract Eq. (9.63)

$$\partial_\mu g_{\nu\lambda} + \partial_\nu g_{\lambda\mu} - \partial_\lambda g_{\mu\nu} = \Gamma^\sigma_{\mu\nu} g_{\sigma\lambda} + \Gamma^\sigma_{\mu\lambda} g_{\sigma\nu} + \Gamma^\sigma_{\nu\lambda} g_{\sigma\mu}$$
$$+ \Gamma^\sigma_{\nu\mu} g_{\sigma\lambda} - \Gamma^\sigma_{\lambda\mu} g_{\sigma\nu} - \Gamma^\sigma_{\lambda\nu} g_{\sigma\mu}$$
$$= 2\Gamma^\sigma_{\mu\nu} g_{\sigma\lambda},$$

or

$$g_{\sigma\lambda} \Gamma^\sigma_{\mu\nu} = \frac{1}{2}[\partial_\mu g_{\nu\lambda} + \partial_\nu g_{\lambda\mu} - \partial_\lambda g_{\mu\nu}]. \tag{9.66}$$

Multiply by $g^{\lambda\rho}$ and use

$$g^{\lambda\rho} g_{\sigma\lambda} = \delta^\rho_\sigma,$$

we get

$$\Gamma^\rho_{\mu\nu} = \frac{1}{2} g^{\lambda\rho}[\partial_\mu g_{\nu\lambda} + \partial_\nu g_{\lambda\mu} - \partial_\lambda g_{\mu\nu}], \tag{9.67}$$

$$\Gamma^\lambda_{\mu\nu} = \frac{1}{2} g^{\lambda\rho}[\partial_\mu g_{\nu\rho} + \partial_\nu g_{\rho\mu} - \partial_\rho g_{\mu\nu}].$$

We end this section with the following remark.

To take into account the curvature, dV^μ is replaced by DV^μ:

$$DV^\mu = (dV^\mu + \Gamma^\mu_{\nu\lambda} V^\nu dx^\lambda), \tag{9.68}$$

or

$$DV^\mu = (\frac{\partial V^\mu}{\partial x^\lambda} + \Gamma^\mu_{\nu\lambda} V^\nu) dx^\lambda$$
$$= V^\mu_{;\lambda} dx^\lambda. \tag{9.69}$$

The term $\Gamma^\mu_{\nu\lambda} dx^\lambda$ arises due to curvature. For parallel transport of vector V^μ along curve $x(s)$ (s is the parameter of curve), $\frac{DV^\mu}{ds} = 0$, instead of $\frac{dV^\mu}{ds} = \frac{\partial V^\mu}{\partial x^\lambda} \frac{dx^\lambda}{ds} = 0$ in flat space.

$$\frac{DV^\mu}{ds} = V^\mu_{;\lambda} \frac{dx^\lambda}{ds}$$
$$= 0. \tag{9.70}$$

For covariant vector

$$DV_\mu = V_{\mu;\lambda} dx^\lambda. \tag{9.71}$$

In particular for tangent $\frac{dx^\mu}{ds}$, for parallel transport

$$\frac{D}{ds}(\frac{dx^\mu}{ds}) = 0. \tag{9.72}$$

Now from Eq. (9.68)

$$\frac{D}{ds}\left(\frac{dx^\mu}{ds}\right) = [\frac{d}{ds}\left(\frac{dx^\mu}{ds}\right) + \Gamma^\mu_{\nu\lambda}\frac{dx^\nu}{ds}\frac{dx^\lambda}{ds}]$$

$$= \frac{d^2x^\mu}{ds^2} + \Gamma^\mu_{\nu\lambda}\frac{dx^\nu}{ds}\frac{dx^\lambda}{ds}. \qquad (9.73)$$

Hence for parallel transport of tangent vector

$$\frac{d^2x^\mu}{ds^2} + \Gamma^\mu_{\nu\lambda}\frac{dx^\nu}{ds}\frac{dx^\lambda}{ds} = 0. \qquad (9.74)$$

This is called geodesic equation.

9.4   Gradient, Curl and Divergence

Gradient:

$$\Phi_{;\lambda} = \frac{\partial\Phi}{\partial x^\lambda} \qquad (9.75)$$

$$= \partial_\lambda\Phi,$$

is covariant vector, Φ being scalar function.

Curl:

The curl of covariant vector V_μ:

$$V_{\mu;\nu} - V_{\nu;\mu} = \partial_\nu V_\mu - \partial_\mu V_\nu, \qquad (9.76)$$

since

$$\Gamma^\lambda_{\mu\nu} = \Gamma^\lambda_{\nu\mu}.$$

Divergence:

$$V^\mu_{;\mu} = \partial_\mu V^\mu + \Gamma^\mu_{\lambda\mu}V^\lambda.$$

Now (cf. Problem 9.6)

$$\Gamma^\mu_{\lambda\mu} = \frac{\partial}{\partial x^\lambda}\ln\sqrt{-g} \qquad (9.77)$$

$$= \partial_\lambda\ln\sqrt{-g},$$

so that divergence

$$V^\mu_{;\mu} = \partial_\mu V^\mu + V^\lambda\partial_\lambda\ln\sqrt{-g}$$

$$= \partial_\mu V^\mu + V^\mu(\frac{1}{\sqrt{-g}}\partial_\mu\sqrt{-g})$$

$$= -\frac{1}{\sqrt{-g}}\partial_\mu(\sqrt{-g}V^\mu). \qquad (9.78)$$

9.5   Problems

9.1 Given

$$F^{\mu\nu} = T^{\mu\nu}_\lambda V^\lambda,$$

show that $T^{\mu\nu}_\lambda$ transforms as mixed tensor of rank 3.

9.2 If f is scalar function show that $\frac{\partial f}{\partial x^\mu}$ transforms as covariant vector and $\frac{\partial^2 f}{\partial x^\mu \partial x^\nu}$ transforms as

$$\frac{\partial^2 f}{\partial \acute{x}^\mu \partial \acute{x}^\nu} = \frac{\partial x^\lambda}{\partial \acute{x}^\mu} \frac{\partial x^\sigma}{\partial \acute{x}^\nu} \frac{\partial^2 f}{\partial x^\lambda \partial x^\sigma} + \frac{\partial^2 x^\sigma}{\partial \acute{x}^\mu \partial \acute{x}^\nu} \frac{\partial f}{\partial x^\sigma},$$

i.e., not as covariant tensor of rank 2.

9.3 Derive Eqs. (9.55)–(9.56) of the text.

9.4 Show that

$$g_{\mu\nu;\lambda} = 0,$$

$$g^{\mu\nu}_{;\lambda} = 0.$$

9.5 Derive the result by using

$$\frac{\partial}{\partial x^\mu} \ln(\det G) = Tr(G^{-1} \frac{\partial}{\partial x^\mu} G).$$

Solution

To derive the above result, we use the relations

$$\det(GF) = \det G \det F,$$

$$\det G = e^{Tr \ln G}.$$

Now

$$\det\left(G + \delta G\right) = \det\left(G\left(1 + G^{-1}\delta G\right)\right)$$

$$= \det G + \det\left(1 + G^{-1}\delta G\right), \qquad (9.79)$$

$$\det\left(1 + G^{-1}\delta G\right) = e^{Tr\left[\ln\left(1 + G^{-1}\delta G\right)\right]},$$

$$\ln\left(1 + G^{-1}\delta G\right) = \sum_n \frac{1}{n!}\left(G^{-1}\delta G\right)^n$$

$$= G^{-1}\delta G + O\left(\delta G^2\right).$$

Thus

$$\det\left(G + \delta G\right) = e^{Tr\left[G^{-1}\delta G\right]} = \left(1 + Tr\left(G^{-1}\delta G\right)\right) + O\left(\delta G^2\right).$$

Therefore, from Eq. (9.79)

$$\det\left(G + \delta G\right) = \det G\left(1 + Tr\left(G^{-1}\delta G\right)\right).$$

Hence

$$\ln \det (G + \delta G) = \ln \det G + \ln \left(1 + Tr \left(G^{-1} \delta G\right)\right)$$

or

$$\delta \ln \det G = Tr \left(G^{-1} \delta G\right).$$

Hence

$$\frac{\partial}{\partial x^\mu} \ln(\det G) = \frac{\partial}{\partial x^\mu} Tr(G^{-1} \frac{\partial}{\partial x^\mu} G) = Tr(G^{-1} \frac{\partial}{\partial x^\mu} G).$$

9.6 Show that

$$\Gamma^\lambda_{\mu\lambda} = \frac{1}{2} g^{\lambda\sigma} \frac{\partial}{\partial x^\mu} g_{\sigma\lambda}$$

$$= \frac{1}{2} (G^{-1} \frac{\partial}{\partial x^\mu} G).$$

Using the result of Problem 9.5 show that

$$\Gamma^\lambda_{\mu\lambda} = \frac{\partial}{\partial x^\mu} \ln \sqrt{-g} = \frac{1}{\sqrt{-g}} \frac{\partial}{\partial x^\mu} \sqrt{-g}.$$

Solution

$$\Gamma^\lambda_{\mu\lambda} = \frac{1}{2} g^{\lambda\sigma} \left(\partial_\mu g_{\lambda\sigma} + \partial_\lambda g_{\sigma\mu} - \partial_\sigma g_{\mu\lambda}\right).$$

Also

$$\Gamma^\lambda_{\mu\lambda} = \frac{1}{2} g^{\sigma\lambda} \left(\partial_\mu g_{\sigma\lambda} + \partial_\sigma g_{\lambda\mu} - \partial_\lambda g_{\mu\sigma}\right).$$

Therefore

$$\Gamma^\lambda_{\mu\lambda} = \frac{1}{2} g^{\lambda\sigma} \partial_\mu g_{\lambda\sigma} = \frac{1}{2} Tr \left(G^{-1} \partial_\mu G\right)$$

$$= \frac{1}{2} \partial_\mu \ln (\det G).$$

Now

$$\partial_\mu \ln (\det G) = \partial_\mu \ln (-g).$$

Hence

$$\Gamma^\lambda_{\mu\lambda} = \frac{1}{2} g^{\lambda\sigma} \partial_\mu g_{\lambda\sigma}$$

$$= \partial_\mu \ln (-g)$$

$$= \frac{1}{\sqrt{-g}} \frac{\partial}{\partial x^\mu} \sqrt{-g}.$$

9.7 Show that:

$$\frac{\partial}{\partial x^\sigma}(\sqrt{-g}g^{\lambda\sigma}) + \sqrt{-g}\Gamma^\lambda_{\mu\nu}g^{\mu\nu} = 0.$$

Hint: First show that

$$g^{\mu\nu}\Gamma^\lambda_{\mu\nu} = [-\frac{\partial g^{\lambda\sigma}}{\partial x^\sigma} - \frac{1}{2}g^{\lambda\sigma}g^{\mu\nu}\frac{\partial g^{\mu\nu}}{\partial x^\sigma}],$$

then use

$$\frac{\partial}{\partial x^\nu}(g^{\lambda\sigma}g_{\mu\sigma}) = 0,$$

$$\frac{\partial}{\partial x^\mu}(g^{\lambda\sigma}g_{\nu\sigma}) = 0.$$

9.8 In flat space, for parallel transport of a vector V^μ from one point to another along a curve $x(s)$, we have

$$\frac{d}{ds}V^\mu = \frac{dV^\mu}{dx^\lambda}\frac{dx^\lambda}{ds}$$
$$= 0.$$

In particular for a tangent vector

$$V^\mu = \frac{dx^\mu}{ds},$$

$$\frac{d^2x^\mu}{ds^2} = \frac{dx^\lambda}{ds}\partial_\lambda(\frac{dx^\mu}{ds})$$
$$= 0,$$

gives the geodesic in flat space (straight line). Obtain the geodesic in curved space by replacing ∂_λ by covariant derivative ∇_λ in the above equation.

9.9 Show that

$$\nabla_\lambda\epsilon_{\mu\nu\rho\sigma} = \epsilon_{\mu\nu\rho\sigma;\lambda} = 0.$$

9.10 In Euclidean space: ξ^a: Cartesian coordinates and x^i: General coordinates. Find the metric and inverse metric and transformation matrix for the spherical polar coordinates r, θ, ϕ, find g_{ij}, g^{ij} and Christoffel symbols.

Solution

$$dx^i = \frac{\partial x^i}{\partial \xi^\alpha}d\xi^\alpha,$$

$$dx_i = \frac{\partial \xi^\alpha}{\partial x^i}d\xi_\alpha.$$

The Jacobian of the transformation $(\xi \to x)$

$$J = \left| \frac{\partial x}{\partial \xi} \right| = \varepsilon_{ijk} \frac{\partial x^i}{\partial \xi^1} \frac{\partial x^j}{\partial \xi^2} \frac{\partial x^k}{\partial \xi^3}.$$

The inverse transformation

$$d\xi^\alpha = \frac{\partial \xi^\alpha}{\partial x^i} dx^i,$$

$$d\xi_\alpha = \frac{\partial x^i}{\partial \xi^\alpha} dx_i.$$

The Jacobian of transformation $(x \to \xi)$

$$J^{-1} = \left| \frac{\partial \xi}{\partial x} \right| = \varepsilon_{abc} \frac{\partial \xi^a}{\partial x^1} \frac{\partial \xi^b}{\partial x^2} \frac{\partial \xi^c}{\partial x^3}.$$

The metric tensor

$$g_{ij} = \sum_a \frac{\partial \xi^\alpha}{\partial x^i} \frac{\partial \xi^\alpha}{\partial x^j} = g_{ji},$$

$$g^{ij} = \sum_a \frac{\partial x^i}{\partial \xi^\alpha} \frac{\partial x^j}{\partial \xi^\alpha} = g^{ji},$$

$$\det g_{ij} \equiv g = \det \left(\frac{\partial \xi^\alpha}{\partial x^i} \frac{\partial \xi^\alpha}{\partial x^j} \right).$$

$$= J^{-2}.$$

(i) Spherical polar coordinates

$x^1 = r, \ x^2 = \theta, \ x^3 = \phi$

$$x^1 = \sqrt{(\xi^1)^2 + (\xi^2)^2 + (\xi^3)^2} = r, \quad x^2 = \tan^{-1} \frac{\sqrt{(\xi^1)^2 + (\xi^2)^2}}{\xi^3} = \theta.$$

For orthogonal coordinates $g_{ij} = 0, \ i \neq j$

$$g_{11} = 1 = h_1, \ g_{22} = r^2 = h_2, \ g_{33} = r^2 \sin^2 \theta = h_3,$$

$$\det g = g_{11} g_{22} g_{33} = h_1 h_2 h_3 = r^4 \sin^2 \theta,$$

$$J^{-1} = \left| \frac{\partial \xi}{\partial x} \right| = \sqrt{g} = r^2 \sin \theta.$$

Volume element

$$d\xi^1 d\xi^2 d\xi^3 = J^{-1} dx^1 dx^2 dx^3$$

$$= J^{-1} dr d\theta d\phi$$

$$= r^2 \sin \theta dr d\theta d\phi.$$

(ii) Inverse metric

$$g^{11} = \frac{\left(\xi^1\right)^2 + \left(\xi^2\right)^2 + \left(\xi^3\right)^2}{r^2} = 1 = \frac{1}{h_1},$$

$$g^{22} = \frac{1}{r^4}\left(\left(\xi^1\right)^2 + \left(\xi^2\right)^2 + \left(\xi^3\right)^2\right) = \frac{1}{r^2} = \frac{1}{h_2},$$

$$g^{33} = \frac{1}{\left(\xi^1\right)^2 + \left(\xi^2\right)^2} = \frac{1}{r^2 \sin^2\theta} = \frac{1}{h_3}.$$

Christoffel symbols for spherical polar coordinates

$$\Gamma^i_{ii} = \frac{1}{2h_i}\frac{\partial h_i}{\partial x^i}, \ \Gamma^i_{ij} = \Gamma^i_{ji} = \frac{1}{2h_i}\frac{\partial h_i}{\partial x^j}, \Gamma^i_{jj} = -\frac{1}{2h_i}\frac{\partial h_i}{\partial x^j},$$

$$\Gamma^1_{11} = 0 = \Gamma^2_{22} = \Gamma^3_{33},$$

$$\Gamma^1_{22} = -r = \Gamma^r_{\theta\theta}, \ \Gamma^1_{33} = -r\sin^2\theta = \Gamma^r_{\phi\phi},$$

$$\Gamma^2_{12} = \frac{1}{r} = \Gamma^\theta_{r\theta}, \ \Gamma^3_{13} = \frac{1}{r} = \Gamma^\phi_{r\phi},$$

$$\Gamma^2_{33} = -\sin\theta\cos\theta = \Gamma^\theta_{\phi\phi}, \ \Gamma^3_{23} = \cot\theta = \Gamma^\phi_{\theta\phi}, \ \Gamma^2_{23} = 0 = \Gamma^\theta_{\theta\phi}.$$

9.11 For spherical polar coordinates, calculate the divergence

$$\nabla_\mu V^\mu = \nabla_i V^i = V^i{}_{,i}$$

and

$$\nabla_\mu \nabla^\mu f = \nabla_i \nabla_i f.$$

Chapter 10

GEODESIC AND EQUIVALENCE PRINCIPLE

10.1 Geodesic Equation

The geometry of space-time is described by line element giving distance, ds^2, between two infinitesimally close points

$$ds^2 = g_{\mu\nu}(x)dx^\mu dx^\nu \tag{10.1}$$

called interval in Riemannian geometry. A geodesic can be defined by requirement that it is the shortest curve between a point B and another point A, i.e., we have to find a curve $x^\mu = x^\mu(\tau), 0 \leq \tau \leq 1$ which minimizes the integral $s = \int_B^A ds$. Using the variational principle, we characterize a geodesic by requiring

$$0 = \delta \int_B^A \left(\frac{ds}{d\tau}\right) d\tau = \delta \int_0^1 d\tau [g_{\mu\nu} \frac{dx^\mu}{d\tau} \frac{dx^\nu}{d\tau}]^{1/2}$$

$$= \int_0^1 d\tau \frac{1}{2}[g_{\mu\nu}\dot{x}^\mu \dot{x}^\nu]^{-1/2}[(\delta g_{\mu\nu})\dot{x}^\mu \dot{x}^\nu$$

$$+ g_{\mu\nu}\delta(\dot{x}^\mu)\dot{x}^\nu + g_{\mu\nu}\dot{x}^\mu \delta(\dot{x}^\nu)] \tag{10.2}$$

where $\cdot$ denotes differentiation with respect to τ. Now

$$g_{\mu\nu}\dot{x}^\mu \dot{x}^\nu = g_{\mu\nu}\frac{dx^\mu}{d\tau}\frac{dx^\nu}{d\tau},$$

$$ds^2 = g_{\mu\nu}dx^\mu dx^\nu,$$

$$\left(\frac{ds}{d\tau}\right)^2 = g_{\mu\nu}\dot{x}^\mu \dot{x}^\nu,$$

$$(g_{\mu\nu}\dot{x}^\mu \dot{x}^\nu)^{-1/2} = \frac{d\tau}{ds} \tag{10.3}$$

and

$$\delta(\dot{x}^\mu) = \frac{d}{d\tau}\delta(x^\mu). \tag{10.4}$$

Using Eqs. (10.3) and (10.4), we obtain from Eq. (10.2) $[g_{\mu\nu} = g_{\nu\mu}]$

$$0 = \frac{1}{2}\int_B^A ds[\delta g_{\mu\nu}\frac{dx^\mu}{ds}\frac{dx^\nu}{ds} + 2g_{\mu\nu}\frac{d}{ds}(\delta x^\mu)\frac{dx^\nu}{ds}]. \qquad (10.5)$$

Integrating by parts the last term, using

$$|\delta x^\mu|_B^A = 0,$$

$$\int_B^A ds g_{\mu\nu}\frac{d}{ds}(\delta x^\mu)\frac{dx^\nu}{ds} = g_{\mu\nu}\frac{dx^\nu}{ds}\delta x^\mu\Big|_B^A - \int_B^A \delta x^\mu \frac{d}{ds}\left(g_{\mu\nu}\frac{dx^\nu}{ds}\right)$$

$$= -\int_B^A \delta x^\mu \left[\frac{d}{ds}(g_{\mu\nu})\frac{dx^\nu}{ds} + g_{\mu\nu}\frac{d}{ds}\left(\frac{dx^\nu}{ds}\right)\right]ds. \qquad (10.6)$$

Using

$$\delta g_{\mu\nu} = \frac{\partial g_{\mu\nu}}{\partial x^\lambda}\delta x^\lambda,$$

$$\frac{d}{ds}(g_{\mu\nu}) = \frac{\partial g_{\mu\nu}}{\partial x^\lambda}\frac{dx^\lambda}{ds},$$

we obtain

$$0 = \frac{1}{2}\int_B^A ds\left\{\frac{\partial g_{\mu\nu}}{\partial x^\lambda}\delta x^\lambda\frac{dx^\mu}{ds}\frac{dx^\nu}{ds} - 2\delta x^\mu\left(\frac{\partial g_{\mu\nu}}{\partial x^\lambda}\frac{dx^\lambda}{ds}\frac{dx^\nu}{ds} + g_{\mu\nu}\frac{d^2x^\nu}{ds^2}\right)\right\}. \qquad (10.7)$$

Since δx^λ is arbitrary this implies (changing $\mu \longleftrightarrow \lambda$ in the last term)

$$\frac{\partial g_{\mu\nu}}{\partial x^\lambda}\frac{dx^\lambda}{ds}\frac{dx^\nu}{ds} - 2\frac{\partial g_{\nu\lambda}}{\partial x^\mu}\frac{dx^\mu}{ds}\frac{dx^\nu}{ds} - 2g_{\lambda\nu}\frac{d^2x^\nu}{ds^2} = 0. \qquad (10.8)$$

Changing $\nu \longleftrightarrow \mu$, we have

$$2g_{\mu\lambda}\frac{d^2x^\mu}{ds^2} + \left(\frac{\partial g_{\mu\lambda}}{\partial x^\nu} + \frac{\partial g_{\nu\lambda}}{\partial x^\mu}\right)\frac{dx^\mu}{ds}\frac{dx^\nu}{ds} - \frac{\partial g_{\mu\nu}}{\partial x^\lambda}\frac{dx^\mu}{ds}\frac{dx^\nu}{ds} = 0. \qquad (10.9)$$

By multiplying Eq. (10.9) by $g_{\rho\lambda}$ and using $g^{\rho\lambda}g_{\mu\lambda} = \delta_\mu^\rho$, we obtain the geodesic equation $(\rho \leftrightarrow \lambda)$

$$\frac{d^2x^\lambda}{ds^2} + \frac{1}{2}g^{\lambda\rho}\left(\frac{\partial g_{\mu\rho}}{\partial x^\nu} + \frac{\partial g_{\nu\rho}}{\partial x^\mu} - \frac{\partial g_{\mu\nu}}{\partial x^\rho}\right)\frac{dx^\mu}{ds}\frac{dx^\nu}{ds} = 0 \qquad (10.10)$$

or

$$\frac{d^2x^\lambda}{ds^2} + \{^\lambda_{\mu\nu}\}\frac{dx^\mu}{ds}\frac{dx^\nu}{ds} = 0, \qquad (10.11)$$

where

$$\{^\lambda_{\mu\nu}\} = \frac{1}{2}g^{\lambda\rho}\left(\frac{\partial g_{\mu\rho}}{\partial x^\nu} + \frac{\partial g_{\nu\rho}}{\partial x^\mu} - \frac{\partial g_{\mu\nu}}{\partial x^\rho}\right) = \Gamma^\lambda_{\mu\nu} \tag{10.12}$$

is called the Christoffel symbol.

10.2 Equivalence Principle

According to Newton's second law of motion: force exerted on an object of inertial mass m_i is given by

$$\mathbf{F} = m_i\mathbf{a} \tag{10.13}$$

where $\mathbf{a}$ is the acceleration. According to Newton's law of gravitation, the gravitational force exerted on an object is proportional to gradient of a scalar potential ϕ called the gravitational potential

$$\mathbf{F}_g = -m_g\nabla\phi, \tag{10.14}$$

where constant of proportionality m_g is called the gravitational mass or gravitational charge; a quantity specific to gravitational force. Then from the above two equations, we have

$$m_i\mathbf{a} = -m_g\nabla\phi. \tag{10.15}$$

The weak equivalence principle states

$$m_i = m_g. \tag{10.16}$$

Hence equivalence principle gives

$$\mathbf{a} = -\nabla\phi, \tag{10.17}$$

i.e., acceleration due to gravitational force is independent of material content of the object characterized by the inertial mass m_i. This has been experimentally verified to a high degree of precision in Eötvös experiments. Weak equivalence principle is specific to gravitational force. Contrast it with a force which a charged particle experiences in an electric field $\mathbf{E}$:

$$\mathbf{F} = m_i\mathbf{a} = e\mathbf{E} = -e\nabla\phi \tag{10.18}$$

$$\mathbf{a} = \frac{-e}{m_i}\nabla\phi \tag{10.19}$$

where ϕ is the electrostatic potential. Here $\mathbf{a}$ is no longer independent of material content of the particle.

The gravitational force has another characteristic viz that it may be made to vanish at a space-time point in an appropriate reference frame. Such a reference frame is provided by a falling box in the gravitational field of the earth. To an observer in the box, all bodies that are falling freely will

appear to be at rest and physical events will happen in the box just the
same way as if the box were at rest, in spite of the fact that gravitational
force is acting. This is expressed as Einstein's equivalence principle: there
is a freely falling coordinate system ξ^α in which the equation of motion is
a straight line in space and time

$$\frac{d^2\xi^\alpha}{d\tau^2} = 0 \tag{10.20}$$

τ: proper time

$$d\tau^2 = \eta_{\alpha\beta}d\xi^\alpha d\xi^\beta. \tag{10.21}$$

The transformation from inertial coordinate ξ^α to general coordinate x^μ is
given by

$$d\xi^\alpha = \frac{\partial\xi^\alpha}{\partial x^\mu}dx^\mu, \tag{10.22}$$

$$\frac{d\xi^\alpha}{d\tau} = \frac{\partial\xi^\alpha}{\partial x^\mu}\frac{dx^\mu}{d\tau}. \tag{10.23}$$

Thus

$$d\tau^2 = \eta_{\alpha\beta}\frac{\partial\xi^\alpha}{\partial x^\mu}\frac{\partial\xi^\beta}{\partial x^\nu}dx^\mu dx^\nu = g_{\mu\nu}dx^\mu dx^\nu = ds^2. \tag{10.24}$$

This implies that there exists a coordinate system ξ^μ locally, i.e., an in-
finitely small world region in which gravity has no influence. Thus there
exists a local inertial frame-coordinate centered on a point P in which

$$g_{\mu\nu}(x_P) = \eta_{\mu\nu}.$$

From Eq. (10.20), using Eq. (10.23)

$$0 = \frac{d^2\xi^\alpha}{d\tau^2} = \frac{d}{d\tau}\left(\frac{\partial\xi^\alpha}{\partial x^\mu}\frac{dx^\mu}{d\tau}\right)$$

$$= \frac{\partial\xi^\alpha}{\partial x^\mu}\frac{d^2x^\mu}{d\tau^2} + \frac{dx^\mu}{d\tau}\frac{d}{d\tau}\left(\frac{\partial\xi^\alpha}{\partial x^\mu}\right)$$

$$= \frac{\partial\xi^\alpha}{\partial x^\mu}\frac{d^2x^\mu}{d\tau^2} + \frac{dx^\mu}{d\tau}\frac{\partial^2\xi^\alpha}{\partial x^\mu\partial x^\nu}\frac{dx^\nu}{d\tau}. \tag{10.25}$$

Multiplying both sides by $\frac{\partial x^\lambda}{\partial\xi^\alpha}$, we have

$$0 = \frac{\partial x^\lambda}{\partial\xi^\alpha}\frac{\partial\xi^\alpha}{\partial x^\mu}\frac{d^2x^\mu}{d\tau^2} + \frac{\partial x^\lambda}{\partial\xi^\alpha}\frac{\partial^2\xi^\alpha}{\partial x^\mu\partial x^\nu}\frac{dx^\mu}{d\tau}\frac{dx^\nu}{d\tau}. \tag{10.26}$$

Thus

$$\delta^\lambda_\mu\frac{d^2x^\mu}{d\tau^2} + \frac{\partial x^\lambda}{\partial\xi^\alpha}\frac{\partial^2\xi^\alpha}{\partial x^\mu\partial x^\nu}\frac{dx^\mu}{d\tau}\frac{dx^\nu}{d\tau} = 0.$$

Hence

$$\frac{d^2x^\lambda}{d\tau^2} + \Gamma^\lambda_{\mu\nu}\frac{dx^\mu}{d\tau}\frac{dx^\nu}{d\tau} = 0, \qquad (10.27)$$

where

$$\Gamma^\lambda_{\mu\nu} = \frac{\partial x^\lambda}{\partial \xi^\alpha}\frac{\partial^2\xi^\alpha}{\partial x^\mu \partial x^\nu}$$

is the Christoffel symbol. Here we have derived the geodesic equation from the Einstein's equivalence principle.

This equation should take the same form at point P as in Eq. (10.20), namely,

$$\left.\frac{d^2x^\mu}{d\tau^2}\right|_P = 0 \qquad (10.28)$$

if Christoffel symbols vanish at P, thus we can say that a freely falling frame (in which Christoffel symbols vanish) is a local inertial frame along a geodesic. Equation (10.27) can also be written as (cf. Eq. (10.24))

$$\frac{d^2x^\lambda}{ds^2} + \Gamma^\lambda_{\mu\nu}\frac{dx^\mu}{ds}\frac{dx^\nu}{ds} = 0. \qquad (10.29)$$

10.3 Weak Field and Low Velocity Limit: Gravity as a Metric Phenomenon

In the weak field limit we may write

$$g_{\mu\nu} = \eta_{\mu\nu} + \varepsilon_{\mu\nu}, \qquad |\varepsilon_{\mu\nu}| \ll 1. \qquad (10.30)$$

We also take it to be time-independent. Then the line element

$$\begin{aligned}
ds^2 &= g_{\mu\nu}dx^\mu dx^\nu \\
&= \eta_{\mu\nu}dx^\mu dx^\nu + \varepsilon_{\mu\nu}dx^\mu dx^\nu \\
&= c^2 dt^2 - d\mathbf{x}^2 + \varepsilon_{\mu\nu}dx^\mu dx^\nu. \qquad (10.31)
\end{aligned}$$

We also assume that the velocity of the particle along the geodesic is much less than c. Thus from Eq. (10.31)

$$\begin{aligned}
\left(\frac{ds}{dt}\right)^2 &= c^2 - v^2 + \epsilon_{\mu\nu}\frac{dx^\mu}{dt}\frac{dx^\nu}{dt} \\
&= c^2 - v^2 + \epsilon_{00}\left(\frac{dx^0}{dt}\right)^2 + \mathcal{O}\left(\epsilon\frac{v}{c}\right) \\
&= c^2(1+\epsilon_{00}) + \mathcal{O}\left(\frac{v^2}{c^2}\right) + \mathcal{O}\left(\epsilon\frac{v}{c}\right). \qquad (10.32)
\end{aligned}$$

Thus

$$(\frac{ds}{dt})^2 \approx c^2(1 + \varepsilon_{00}). \tag{10.33}$$

We apply the same approximation to the geodesic equation (10.29)

$$\frac{d^2x^\lambda}{dt^2}(\frac{dt}{ds})^2 + \Gamma^\lambda_{\mu\nu}\frac{dx^\mu}{dt}\frac{dx^\nu}{dt}\left(\frac{dt}{ds}\right)^2 = 0. \tag{10.34}$$

We note that $\Gamma^\lambda_{\mu\nu}$ is already of order ϵ, thus we obtain to the order we are working (Problem 10.2)

$$\frac{1}{c^2(1 + \varepsilon_{00})}\{\frac{d^2x^\lambda}{dt^2} + c^2\Gamma^\lambda_{00}\} = 0, \tag{10.35}$$

where

$$\Gamma^\lambda_{00} = \frac{1}{2}\eta^{\lambda\rho}[2\partial_0\varepsilon_{0\rho} - \partial_\rho\varepsilon_{00}]. \tag{10.36}$$

$$\Gamma^i_{00} = \frac{1}{2}\partial_i\varepsilon_{00}. \tag{10.37}$$

$$\Gamma^0_{00} = \frac{1}{2}\partial_0\varepsilon_{00} = 0, \tag{10.38}$$

since ε_{00} is independent of time. Thus we obtain from Eq. (10.35)

$$\frac{d^2x^i}{dt^2} = -\frac{c^2}{2}\partial_i\varepsilon_{00}. \tag{10.39}$$

This is identical with the Newton's equation of motion in a classical gravitational field derived for the scalar potential Φ:

$$\frac{d^2\mathbf{x}}{dt^2} = -\nabla\Phi \tag{10.40}$$

if we identify $\frac{c^2}{2}\varepsilon_{00}$ with the gravitational potential Φ:

$$\varepsilon_{00} = \frac{2}{c^2}\Phi \tag{10.41}$$

and

$$g_{00} \approx 1 + \frac{2\Phi}{c^2}. \tag{10.42}$$

The other components do not enter into our approximation except that they are time independent and satisfy (10.30).

To summarize in the weak field ($|\varepsilon_{\mu\nu}| \ll 1$) and low velocity ($\frac{v}{c} \ll 1$) limit, the geodesic equation representing a purely geometric relation is equivalent to a purely mechanistic relation represented by Newton's equation provided that g_{00} satisfies the relation (10.42).

The above result can also be used to determine the influence of gravitational field on clocks. This is determined by g_{00}. The interval ds between two infinitesimally separated events occurring at the same point in space is just the proper time $cd\tau$:

$$ds = cd\tau$$

$$\tau = \frac{1}{c} \int ds$$

$$= \frac{1}{c} \int \sqrt{g_{00}} dx^0$$

$$= \frac{x^0}{c} \sqrt{1 + \frac{2\Phi}{c^2}} = \frac{x^0}{c}(1 + \frac{\Phi}{c^2}). \tag{10.43}$$

Thus

$$t \approx \tau(1 - \frac{\Phi}{c^2}). \tag{10.44}$$

This means that if one of the two clocks is placed in a gravitational field, the clock which has been in gravitational field will thereafter be slower (Φ is negative).

This can be used to calculate the gravitational redshift of spectral lines. The frequency ν_0 expressed in terms of the world line x_0/c remains constant during the propagation of light ray. It is the frequency ν associated with proper time which is different at different points of space. From Eq. (10.44), we have

$$\frac{1}{\nu_0} = \frac{1}{\nu} \left(1 - \frac{\Phi}{c^2} \right)$$

$$\nu \approx \nu_0(1 - \frac{\Phi}{c^2}). \tag{10.45}$$

This means that the light frequency increases with increasing absolute value of gravitational potential Φ., i.e., as we approach the bodies producing the field; conversely as the light reaches the body, the frequency decreases. Suppose the light ray emitted at a point in the sun, where the gravitational potential is Φ, has the frequency ν; then its frequency at a point on the earth where the gravitational potential is Φ_E is given by

$$\nu_E = \nu_0(1 - \frac{\Phi_E}{c^2}) = \nu \frac{1 - \frac{\Phi_E}{c^2}}{1 - \frac{\Phi_S}{c^2}}$$

$$= \nu \left(1 + \frac{\Phi_S - \Phi_E}{c^2} \right). \tag{10.46}$$

Thus

$$\frac{\Delta \nu}{\nu} = \frac{\nu_E - \nu}{\nu} = \frac{\Phi_S - \Phi_E}{c^2} = \frac{|\Phi_E| - |\Phi_S|}{c^2} \qquad (10.47)$$

since $|\Phi_S| > |\Phi_E|$, we have the "redshift".

10.4   Problems

10.1 Derive Eqs. (10.36)–(10.39) of the text.

10.2 Show that the geodesic equation

$$\frac{d^2 x^\lambda}{d\tau^2} + \Gamma^\lambda_{\mu\nu} \frac{dx^\mu}{d\tau} \frac{dx^\nu}{d\tau} = 0,$$

can be expressed as

$$p^\mu \nabla_\mu p^\lambda \equiv p^\mu [\partial_\mu p^\lambda + \Gamma^\lambda_{\mu\nu} p^\nu] = 0,$$

$$p^\mu \nabla_\mu p_\lambda \equiv p^\mu [\partial_\mu p_\lambda - \Gamma^\nu_{\mu\lambda} p_\nu] = 0,$$

where

$$p^\lambda = m U^\lambda = m \frac{dx^\lambda}{d\tau}.$$

Show that

$$\frac{dU_\lambda}{d\tau} = \frac{1}{2} (\partial_\lambda g_{\mu\nu}) U^\mu U^\nu.$$

Solution: Geodesic equation in terms of p^λ

$$\frac{dp^\lambda}{d\tau} + \Gamma^\lambda_{\mu\nu} p^\mu p^\nu = 0. \qquad (10.48)$$

Then using

$$\frac{d}{d\tau}(p^\lambda) = \frac{\partial p^\lambda}{\partial x^\mu} \frac{dx^\mu}{d\tau}, \qquad (10.49)$$

we get from Eq. (10.48)

$$[\frac{\partial p^\lambda}{\partial x^\mu} + \Gamma^\lambda_{\mu\nu} p^\nu] \frac{1}{m} p^\mu,$$

$$p^\mu \nabla_\mu p^\lambda = 0.$$

Using

$$p^\lambda = g^{\lambda\rho} p_\rho,$$

and

$$\nabla_\mu g^{\lambda\rho} = 0,$$

$$g^{\lambda\rho} p^\mu \nabla_\mu p_\rho = g^{\rho\lambda} p^\mu \nabla_\mu p_\lambda = 0.$$

Hence

$$p^\mu \nabla_\mu p_\lambda \equiv p^\mu [\frac{\partial p_\lambda}{\partial x^\mu} - \Gamma^\rho_{\mu\lambda} p_\rho] = 0. \tag{10.50}$$

From Eqs. (10.49) and (10.50)

$$m\frac{dp_\lambda}{d\tau} = \Gamma^\rho_{\mu\lambda} p^\mu p_\rho. \tag{10.51}$$

Using

$$\Gamma^\rho_{\mu\lambda} p^\mu p_\rho = \frac{1}{2}\frac{\partial g_{\mu\sigma}}{\partial x^\lambda} p^\mu p^\sigma,$$

we get from Eq. (10.51)

$$\frac{dU_\lambda}{d\tau} = \frac{1}{2}[\partial_\lambda g_{\mu\nu}]U^\mu U^\nu. \tag{10.52}$$

10.3 Consider a rotating frame, rotating with constant angular velocity ω. In polar coordinates (t, r, θ, z), the metric is given by

$$ds^2 = c^2 d\tau^2 - (dr^2 + r^2 d\theta^2 + dz^2).$$

In rotating frame: $(t', r, \theta', z), \theta = \theta' + \omega t', t' = t$. Show that

$$\frac{dp_r}{d\tau} = -\frac{mr\omega^2}{1 - \frac{r^2\omega^2}{c^2}},$$

implies

$$m\frac{d^2r}{dt^2} = -mr\omega^2.$$

Chapter 11

CURVATURE TENSOR AND EINSTEIN'S FIELD EQUATIONS

11.1 Curvature Tensor

The metric tensor $g_{\mu\nu}$ and the Christoffel symbol $\Gamma^{\rho}_{\mu\lambda}$ are the characteristic of curvature. As is well known in flat space one can introduce curvilinear coordinates with metric tensor and Christoffel symbols characteristic of curvature of coordinates. The curvature tensor or Riemannian tensor vanishes for the curvilinear coordinates describing flat space. Thus it is the curvature which determines the geometry; it vanishes for the Minskowski space.

A covariant derivative, or connection ∇_{μ} has an associated tensor field $R^{\rho}_{\mu\nu\lambda}$, called Riemannian curvature tensor:

$$R^{\rho}_{\mu\nu\lambda} = \partial_{\nu}\Gamma^{\rho}_{\mu\lambda} - \partial_{\lambda}\Gamma^{\rho}_{\mu\nu} + \Gamma^{\sigma}_{\mu\lambda}\Gamma^{\rho}_{\sigma\nu} - \Gamma^{\sigma}_{\mu\nu}\Gamma^{\rho}_{\sigma\lambda}. \tag{11.1}$$

Define

$$V_{\mu;\nu\lambda} = (V_{\mu;\nu})_{;\lambda}. \tag{11.2}$$

One can show that (see Problem 11.1)

$$V_{\mu;\nu\lambda} - V_{\mu;\lambda\nu} = R^{\rho}_{\mu\nu\lambda}V_{\rho}. \tag{11.3}$$

Since the left-hand side of Eq. (11.3) is a covariant tensor of rank 3, V_{ρ} is a covariant vector, it follows from quotient rule that $R^{\rho}_{\mu\nu\lambda}$ must be a tensor of rank 4. The curvature tensor is then defined as

$$R_{\tau\mu\nu\lambda} = g_{\tau\rho}R^{\rho}_{\mu\nu\lambda} \tag{11.4}$$

$R^{\rho}_{\mu\nu\lambda}$ is antisymmetric in the last two indices. Now

$$\Gamma^{\rho}_{\mu\lambda} = \frac{1}{2}g^{\rho\sigma}[\partial_{\lambda}g_{\mu\sigma} + \partial_{\mu}g_{\lambda\sigma} - \partial_{\sigma}g_{\mu\nu}] \tag{11.5}$$

so that

$$g_{\tau\rho}\Gamma^\rho_{\mu\lambda} = \frac{1}{2}[\partial_\lambda g_{\mu\tau} + \partial_\mu g_{\lambda\tau} - \partial_\tau g_{\mu\lambda}] \tag{11.6}$$

and

$$\partial_\nu(g_{\tau\rho}\Gamma^\rho_{\mu\lambda}) = \frac{1}{2}[\partial_\nu\partial_\lambda g_{\mu\tau} + \partial_\nu\partial_\mu g_{\lambda\tau} - \partial_\nu\partial_\tau g_{\mu\lambda}]. \tag{11.7}$$

Now

$$\partial_\nu(g_{\tau\rho}\Gamma^\rho_{\mu\lambda}) = (\partial_\nu g_{\tau\rho})\Gamma^\rho_{\mu\lambda} + g_{\tau\rho}\partial_\nu\Gamma^\rho_{\mu\lambda} \tag{11.8}$$

and using

$$g_{\tau\rho;\nu} = 0, \tag{11.9}$$

i.e.,

$$\partial_\nu g_{\tau\rho} - \Gamma^\sigma_{\tau\nu}g_{\sigma\rho} - \Gamma^\sigma_{\rho\nu}g_{\tau\sigma} = 0 \tag{11.10}$$

we obtain from Eqs. (11.8) and (11.10)

$$\partial_\nu\left(g_{\tau\rho}\Gamma^\rho_{\mu\lambda}\right) = \left(\Gamma^\sigma_{\tau\nu}g_{\sigma\rho} + \Gamma^\sigma_{\rho\nu}g_{\tau\sigma}\right)\Gamma^\rho_{\mu\lambda} + g_{\tau\rho}\partial_\nu\Gamma^\rho_{\mu\lambda}. \tag{11.11}$$

Thus from Eqs. (11.7) and (11.11)

$$g_{\tau\rho}\partial_\nu\Gamma^\rho_{\mu\lambda} = \frac{1}{2}[\partial_\nu\partial_\lambda g_{\mu\tau} + \partial_\nu\partial_\mu g_{\lambda\tau} - \partial_\nu\partial_\tau g_{\mu\lambda}] - \Gamma^\sigma_{\tau\nu}\Gamma^\rho_{\mu\lambda}g_{\sigma\rho} - \Gamma^\sigma_{\rho\nu}\Gamma^\rho_{\mu\lambda}g_{\tau\sigma}. \tag{11.12}$$

Using this equation and similar one for $g_{\tau\rho}\partial_\lambda\Gamma^\rho_{\mu\nu}$, we obtain from Eq. (11.4)

$$R_{\tau\mu\nu\lambda} = g_{\tau\rho}\left(\partial_\nu\Gamma^\rho_{\mu\lambda} - \partial_\lambda\Gamma^\rho_{\mu\nu}\right) + g_{\tau\rho}\left(\Gamma^\sigma_{\mu\lambda}\Gamma^\rho_{\sigma\nu} - \Gamma^\sigma_{\mu\nu}\Gamma^\rho_{\sigma\lambda}\right),$$

$$R_{\tau\mu\nu\lambda} = \frac{1}{2}(\partial_\lambda\partial_\tau g_{\mu\nu} - \partial_\lambda\partial_\mu g_{\tau\nu} + \partial_\nu\partial_\mu g_{\tau\lambda} - \partial_\nu\partial_\tau g_{\mu\lambda}) + g_{\rho\sigma}(\Gamma^\rho_{\mu\nu}\Gamma^\sigma_{\lambda\tau} - \Gamma^\rho_{\mu\lambda}\Gamma^\sigma_{\nu\tau}). \tag{11.13}$$

Symmetry properties of $R_{\tau\mu\nu\lambda}$: (follow from Eq. (11.13))

 (A) Antisymmetries

$$R_{\tau\mu\nu\lambda} = -R_{\mu\tau\nu\lambda} \tag{11.14}$$

$$R_{\tau\mu\nu\lambda} = -R_{\tau\mu\lambda\nu} \tag{11.15}$$

 (B) cyclicity (in $\mu\nu\lambda$)

$$R_{\tau\mu\nu\lambda} + R_{\tau\nu\lambda\mu} + R_{\tau\lambda\mu\nu} = 0 \tag{11.16}$$

 (C) symmetry

$$R_{\tau\mu\nu\lambda} = R_{\nu\lambda\tau\mu}. \tag{11.17}$$

Thus out of $4^4 = 256$ components, all are not independent. To count independent components we consider a general spacetime of N dimensions. We see that the curvature term is antisymmetric in each of the index pair $A = \tau\mu$ and $B = \nu\lambda$, but is symmetric under interchanging of pairs A and B. Thus each of these pairs can assume $\frac{N^2-N}{2} = \frac{1}{2}N(N-1)$ values. Hence treating each of these pairs A and B as one index capable of taking $\frac{1}{2}N(N-1)$ values, the symmetry under the interchange of A and B implies that we have the same number of components as those of $\frac{1}{2}N(N-1) \times \frac{1}{2}N(N-1)$ symmetric matrix. Thus for $N = 4$, we have 6×6 symmetric matrix which has 21 independent components. The cyclicity adds one more constraint so that curvature tensor has 20 independent components. For N dimensions, the number of independent components of curvature tensor C_N:

$$C_N = \frac{1}{12}N^2\left(N^2 - 1\right).$$

First we note that the Riemannian tensor is antisymmetric in the last two indices

$$R^\rho_{\mu\nu\lambda} = -R^\rho_{\mu\lambda\nu}. \tag{11.18}$$

We now define the following derived curvature tensor:

1) Ricci Tensor:

$$R_{\mu\nu} = R^\lambda_{\mu\lambda\nu} = R_{\tau\mu\lambda\nu}g^{\tau\lambda}. \tag{11.19}$$

Using the symmetry property (C) and $g^{\tau\lambda} = g^{\lambda\tau}$

$$R_{\mu\nu} = R_{\lambda\nu\tau\mu}g^{\lambda\tau} = R^\tau_{\nu\tau\mu} = R_{\nu\mu}, \tag{11.20}$$

i.e., it is symmetric. Thus it has 10 independent components. Written in full

$$R_{\mu\nu} = R^\lambda_{\mu\lambda\nu} = \partial_\lambda\Gamma^\lambda_{\mu\nu} - \partial_\nu\Gamma^\lambda_{\mu\lambda} + \Gamma^\rho_{\mu\nu}\Gamma^\lambda_{\rho\lambda} - \Gamma^\rho_{\mu\lambda}\Gamma^\lambda_{\rho\nu} \tag{11.21}$$

2) Ricci Scalar:

$$R = g^{\mu\nu}R_{\mu\nu} = R^\mu_\mu \tag{11.22}$$

3) Einstein tensor:

$$G_{\mu\nu} = R_{\mu\nu} - \frac{1}{2}Rg_{\mu\nu} \tag{11.23}$$

It is symmetric as $R_{\mu\nu}$ and $g_{\mu\nu}$ are symmetric

4) Bianchi identities:

$$R_{\tau\mu\nu\lambda;\sigma} + R_{\tau\mu\sigma\nu;\lambda} + R_{\tau\mu\lambda\sigma;\nu} = 0. \tag{11.24}$$

These can be derived at a given point x, by adopting a locally inertial coordinate system in which $\Gamma^\rho_{\mu\nu}$(but not its derivative) vanish at x so that from Eq. (11.13)

$$R_{\tau\mu\nu\lambda;\sigma} = \frac{1}{2}[\partial_\sigma\partial_\lambda\partial_\tau g_{\mu\nu} - \partial_\sigma\partial_\lambda\partial_\mu g_{\tau\nu} + \partial_\sigma\partial_\nu\partial_\mu g_{\tau\lambda} - \partial_\sigma\partial_\nu\partial_\tau g_{\mu\lambda}]. \quad (11.25)$$

A cyclic permutation of $\nu\lambda\sigma$ gives the Bianchi identities. The equations are manifestly covariant, so since they hold in locally inertial systems, they hold in general.

Rewriting Eq. (11.24):

$$\nabla_\sigma R_{\tau\mu\nu\lambda} + \nabla_\lambda R_{\sigma\nu\tau\mu} + \nabla_\nu R_{\tau\mu\lambda\sigma} = 0. \quad (11.26)$$

Note in rewriting the above form, we have used the symmetry property (C). Multiplying Eq. (11.26) by $g^{\nu\tau}g^{\lambda\mu}$; since

$$\nabla_\sigma g^{\lambda\mu} = \nabla_\lambda g^{\nu\tau} = \nabla_\nu g^{\lambda\mu} = 0, \quad (11.27)$$

we get

$$g^{\nu\tau}\nabla_\sigma g^{\lambda\mu} R_{\tau\mu\nu\lambda} + g^{\lambda\mu}\nabla_\lambda g^{\nu\tau} R_{\sigma\nu\tau\mu} + g^{\nu\tau}\nabla_\nu g^{\lambda\mu} R_{\tau\mu\lambda\sigma} = 0. \quad (11.28)$$

Thus

$$g^{\nu\tau}\nabla_\sigma R^\lambda_{\tau\nu\lambda} + g^{\lambda\mu}\nabla_\lambda R^\tau_{\sigma\tau\mu} + g^{\nu\tau}\nabla_\nu R^\lambda_{\tau\lambda\sigma} = 0. \quad (11.29)$$

Hence we have

$$-g^{\nu\tau}\nabla_\sigma R_{\tau\nu} + g^{\lambda\mu}\nabla_\lambda R_{\sigma\mu} + g^{\nu\tau}\nabla_\nu R_{\tau\sigma} = 0 \quad (11.30)$$

or

$$-\nabla_\sigma R^\nu_\nu + \nabla^\mu R_{\sigma\mu} + \nabla^\tau R_{\tau\sigma} = 0.$$

Thus finally (changing $\sigma \to \nu$):

$$\nabla^\mu R_{\mu\nu} - \frac{1}{2}\nabla_\nu R = 0. \quad (11.31)$$

This implies

$$\nabla^\mu\left(R_{\mu\nu} - \frac{1}{2}g_{\mu\nu}R\right) = 0. \quad (11.32)$$

Hence for Einstein tensor:

$$G_{\mu\nu} = R_{\mu\nu} - \frac{1}{2}g_{\mu\nu}R \quad (11.33)$$

$$G^{\mu\nu} = R^{\mu\nu} - \frac{1}{2}g^{\mu\nu}R$$

we have

$$\nabla^\mu G_{\mu\nu} = 0$$

or equivalently

$$\nabla_\mu G^{\mu\nu} = 0. \tag{11.34}$$

Now $R_{\tau\mu\nu\lambda}$ has 20 components-10 of which are given by symmetric tensor $R_{\mu\nu}$ and the remaining 10 by a tensor called the Weyl tensor $C_{\tau\mu\nu\lambda}$ which for 4 dimensions is:

$$C_{\tau\mu\nu\lambda} = R_{\tau\mu\nu\lambda} + \frac{1}{2}(g_{\tau\lambda}R_{\mu\nu} + g_{\mu\nu}R_{\tau\lambda} - g_{\tau\nu}R_{\mu\lambda} - g_{\mu\lambda}R_{\tau\nu}) + \frac{1}{6}(g_{\tau\nu}g_{\mu\lambda} - g_{\tau\lambda}g_{\mu\nu})R. \tag{11.35}$$

Weyl tensor has the same symmetries as $R_{\tau\mu\nu\lambda}$, but also satisfies (multiplying by $g^{\nu\tau}$)

$$C^\nu_{\mu\nu\lambda} = R_{\mu\lambda} + \frac{1}{2}(R_{\mu\lambda} + R_{\mu\lambda} - 4R_{\mu\lambda} - g_{\mu\lambda}R) + \frac{1}{6}(4g_{\mu\lambda} - g_{\mu\lambda})R = 0. \tag{11.36}$$

Thus

$$R_{\tau\mu\nu\lambda} = C_{\tau\mu\nu\lambda} \text{ if } R_{\tau\mu} = 0.$$

Now in the vacuum region of space-time $R_{\tau\mu} = 0$ (see below), it is the Weyl tensor which describes a gravitational field in vacuum. Noting that Weyl tensor makes its appearance in a space-time dimension four, implying that if we lived in a universe where space-time had only 2 or 3 dimensions, gravity would not exist in the vacuum region according to general relativity.

11.2 Einstein's Field Equations

As gravitational fields reside in the curvature of space-time, further presence of matter causes fabric of space-time to warp, i.e., produce space-time curvature. Thus we expect some sort of relationship between curvature tensor $R_{\tau\mu\nu\lambda}$ and $T_{\mu\nu}$, the energy momentum tensor, which for a scalar field φ is

$$T^{\mu\nu} = \partial^\mu\varphi\partial^\nu\varphi - \frac{1}{2}g^{\mu\nu}(\partial_\lambda\varphi\partial^\lambda\varphi - m^2\varphi^2). \tag{11.37}$$

Now we can construct a tensor out of $R_{\tau\mu\nu\lambda}$ having the same symmetries as $T^{\mu\nu}$. There are two candidates $R_{\mu\nu}$ and $G_{\mu\nu}$ but since

$$\nabla_\mu T^{\mu\nu} = 0 \tag{11.38}$$

while

$$\nabla_\mu R^{\mu\nu} \neq 0, \; \nabla_\mu G^{\mu\nu} = 0 \tag{11.39}$$

in general, we can discard $R^{\mu\nu}$. Thus in view of Eq. (11.34) we are led to consider the equation

$$G^{\mu\nu} = \kappa T^{\mu\nu} \tag{11.40}$$

where $T^{\mu\nu}$ is stress energy tensor and κ is fixed from the experiment. We will show that it reduces to Newtonian's equation of gravity under condition where we expect Newtonian gravity to hold, if

$$\kappa = \frac{8\pi G}{c^4} \tag{11.41}$$

where G is the Newton's gravitational constant. In summary the simplest relationship between curvature and matter consistent with Newtonian gravity is

$$G^{\mu\nu} = \frac{8\pi G}{c^4} T^{\mu\nu} \tag{11.42}$$

or

$$\left(R^{\mu\nu} - \frac{1}{2} g^{\mu\nu} R\right) = \frac{8\pi G}{c^4} T^{\mu\nu}. \tag{11.43}$$

Note that the R.H.S. of Eq. (11.42) or (11.43) is the measure of matter energy density and the L.H.S. is the measure of the curvature. This is known as Einstein's field equation. An alternative form is sometimes useful. Contraction with $g_{\mu\nu}$ gives

$$R - 2R = \frac{8\pi G}{c^4} T^\mu_\mu. \tag{11.44}$$

Thus the above equation can be written as

$$R^{\mu\nu} = \frac{8\pi G}{c^4} \left(T^{\mu\nu} - \frac{1}{2} g^{\mu\nu} T^\lambda_\lambda\right). \tag{11.45}$$

In vacuum region of spacetime $T^{\mu\nu}$ vanishes so that Einstein's equations in vacuum are

$$R^{\mu\nu} = 0. \tag{11.46}$$

Now in the absence of matter, spacetime is homogeneous and isotropic but not necessarily flat. Since $\nabla_\mu g^{\mu\nu} = 0$ and $\nabla_\mu T^{\mu\nu} = 0$, one can add a term $\Lambda g^{\mu\nu}$ on the R.H.S. of Eq. (11.42)

$$G^{\mu\nu} = \kappa T^{\mu\nu} - \Lambda g^{\mu\nu} \tag{11.47}$$

where Λ is a new universal constant, known as cosmological constant. This equation is also consistent with Newtonian gravity, but only if Λ is chosen to be very small. If we include the cosmological constant, Einstein's field equations become

$$R^{\mu\nu} = \frac{8\pi G}{c^4}(T^{\mu\nu} - \frac{1}{2}g^{\mu\nu}T^\lambda_\lambda) - \Lambda g^{\mu\nu} \tag{11.48}$$

$$R_{\mu\nu} = \frac{8\pi G}{c^4}(T_{\mu\nu} - \frac{1}{2}g_{\mu\nu}T^\lambda_\lambda) - \Lambda g_{\mu\nu}.$$

Now in the absence of matter, Einstein's equations in vacuum are

$$R_{\mu\nu} = -\Lambda g_{\mu\nu}. \tag{11.49}$$

11.3   Newtonian Limit of Field Equations

In free space, neglecting Λ, Einstein's field equations are

$$R_{\mu\nu} = 0, \tag{11.50}$$

i.e.,

$$\partial_\lambda \Gamma^\lambda_{\mu\nu} - \partial_\nu \Gamma^\lambda_{\mu\lambda} + \Gamma^\rho_{\mu\nu}\Gamma^\lambda_{\rho\lambda} - \Gamma^\rho_{\mu\lambda}\Gamma^\lambda_{\rho\nu} = 0, \tag{11.51}$$

$$\Gamma^\rho_{\mu\nu} = \frac{1}{2}g^{\rho\sigma}[\partial_\nu g_{\mu\sigma} + \partial_\mu g_{\nu\sigma} - \partial_\sigma g_{\mu\nu}]. \tag{11.52}$$

In the weak field limit, we may write

$$g_{\mu\nu} = \eta_{\mu\nu} + \epsilon_{\mu\nu}, |\epsilon_{\mu\nu}| \ll 1, \tag{11.53}$$

$$\eta_{\mu\nu} = (1, -1, -1, -1). \tag{11.54}$$

Therefore

$$\Gamma^\rho_{\mu\nu} = \frac{1}{2}\eta^{\rho\sigma}(\partial_\nu \epsilon_{\mu\sigma} + \partial_\mu \epsilon_{\nu\sigma} - \partial_\sigma \epsilon_{\mu\nu}) + \mathcal{O}(\epsilon^2). \tag{11.55}$$

Therefore to first order in ϵ

$$R_{\mu\nu} = \partial_\lambda \Gamma^\lambda_{\mu\nu} - \partial_\nu \Gamma^\lambda_{\mu\lambda}$$

$$= \frac{1}{2}\eta^{\lambda\sigma}[\partial_\lambda(\partial_\nu \epsilon_{\mu\sigma} + \partial_\mu \epsilon_{\nu\sigma} - \partial_\sigma \epsilon_{\mu\nu}) - \partial_\nu(\partial_\lambda \epsilon_{\mu\sigma} + \partial_\mu \epsilon_{\lambda\sigma} - \partial_\sigma \epsilon_{\mu\lambda})]$$

$$= \frac{1}{2}[\partial^\sigma \partial_\mu \epsilon_{\nu\sigma} - \partial^\sigma \partial_\sigma \epsilon_{\mu\nu} - \partial_\nu \partial_\mu \epsilon^\lambda_\lambda + \partial^\lambda \partial_\nu \epsilon_{\mu\lambda}], \tag{11.56}$$

$R_{\mu\nu} = 0$ gives

$$\frac{1}{2}[-\partial^2 \epsilon_{\mu\nu} + \partial^\sigma \partial_\mu \epsilon_{\nu\sigma} + \partial^\sigma \partial_\nu \epsilon_{\mu\sigma} - \partial_\nu \partial_\mu \epsilon^\lambda_\lambda] = 0. \tag{11.57}$$

Thus

$$R_{00} = \frac{1}{2}\left[-\partial^2\epsilon_{00} + 2\partial^\sigma\partial_0\epsilon_{0\sigma} - \partial_0^2\epsilon_\lambda^\lambda\right] = 0. \tag{11.58}$$

Hence, for static field, this gives

$$\nabla^2\frac{\epsilon_{00}}{2} = 0 \tag{11.59}$$

or since as already seen, $\epsilon_{00} = \frac{2}{c^2}\Phi$, we have the Laplace equation

$$\nabla^2\Phi = 0.$$

Thus the Einstein equation in vacuum reduces to the Newtonian equation in vacuum

$$\nabla^2\Phi = 0.$$

In the presence of a point mass source, the equation is replaced by (11.48)

$$\frac{1}{2}\nabla^2\epsilon_{00} = \kappa\left(T^{00} - \frac{1}{2}(1 + \epsilon_{00})T_\mu^\mu\right) \approx \kappa\left(T^{00} - \frac{1}{2}T_0^0\right)$$

$$\approx \frac{c^2}{2}\kappa m\delta^3(r). \tag{11.60}$$

Thus

$$\nabla^2\Phi = c^4\frac{\kappa}{2}m\delta^3(r) \tag{11.61}$$

which has solution

$$\Phi(r) = -\kappa\frac{mc^4}{8\pi r}. \tag{11.62}$$

Comparing it with the Newtonian potential due to mass m

$$\Phi(r) = -\frac{Gm}{r} \tag{11.63}$$

we obtain

$$\kappa = \frac{8\pi G}{c^4}. \tag{11.64}$$

11.4   Problems

11.1 Show that

$$V_{\mu;\nu;\lambda} - V_{\mu;\lambda;\nu} = R_{\mu\nu\lambda}^\rho V_\rho.$$

Solution

$$V_{\mu;\nu} = \partial_\nu V_\mu - \Gamma^\sigma_{\mu\nu} V_\sigma,$$

$$\Gamma^\sigma_{\mu\nu} = \Gamma^\sigma_{\nu\mu},$$

$$\Gamma^\sigma_{\mu\lambda} = \Gamma^\sigma_{\lambda\mu}.$$

Thus

$$V_{\mu;\nu;\lambda} = \partial_\lambda(V_{\mu;\nu}) - \Gamma^\rho_{\lambda\mu} V_{\rho;\nu} - \Gamma^\rho_{\lambda\nu} V_{\mu;\rho},$$

since

$$(V_{\mu;\nu})_{;\lambda} - (V_{\mu;\lambda})_{;\nu}$$

$$= \partial_\lambda[\partial_\nu V_\mu - \Gamma^\sigma_{\mu\nu} V_\sigma] - \partial_\nu[\partial_\lambda V_\mu - \Gamma^\sigma_{\mu\lambda} V_\sigma]$$

$$\quad - \Gamma^\rho_{\lambda\mu}(\partial_\nu V_\rho - \Gamma^\tau_{\rho\nu} V_\tau) - \Gamma^\rho_{\lambda\nu}(\partial_\rho V_\mu - \Gamma^\tau_{\mu\rho} V_\tau)$$

$$\quad + \Gamma^\rho_{\nu\mu}(\partial_\lambda V_\rho - \Gamma^\tau_{\rho\lambda} V_\tau) + \Gamma^\rho_{\nu\lambda}(\partial_\rho V_\mu - \Gamma^\tau_{\mu\rho} V_\tau)$$

$$= -\partial_\lambda(\Gamma^\sigma_{\mu\nu} V_\sigma) + \partial_\nu(\Gamma^\sigma_{\mu\lambda} V_\sigma) - \Gamma^\rho_{\lambda\mu}\partial_\nu V_\rho + \Gamma^\rho_{\lambda\mu}\Gamma^\tau_{\rho\nu} V_\tau$$

$$\quad + \Gamma^\rho_{\nu\mu}\partial_\lambda V_\rho - \Gamma^\rho_{\nu\mu}\Gamma^\tau_{\rho\lambda} V_\tau$$

$$= (-\partial_\lambda\Gamma^\sigma_{\mu\nu}) V_\sigma - \Gamma^\sigma_{\mu\nu}\partial_\lambda V_\rho + (\partial_\nu\Gamma^\sigma_{\mu\lambda}) V_\sigma + \Gamma^\sigma_{\mu\lambda}\partial_\nu V_\sigma - \Gamma^\rho_{\lambda\mu}\partial_\nu V_\rho + \Gamma^\rho_{\lambda\mu}\Gamma^\tau_{\rho\nu} V_\tau$$

$$\quad + \Gamma^\rho_{\nu\mu}\partial_\lambda V_\rho - \Gamma^\rho_{\nu\mu}\Gamma^\tau_{\rho\lambda} V_\tau$$

$$= (-\partial_\lambda\Gamma^\sigma_{\mu\nu} + \partial_\nu\Gamma^\sigma_{\mu\lambda}) V_\sigma + \Gamma^\rho_{\lambda\mu}\Gamma^\tau_{\rho\nu} V_\tau - \Gamma^\rho_{\nu\mu}\Gamma^\tau_{\rho\lambda} V_\tau$$

$$= (\partial_\nu\Gamma^\sigma_{\mu\lambda} - \partial_\lambda\Gamma^\sigma_{\mu\nu} + \Gamma^\rho_{\mu\lambda}\Gamma^\sigma_{\rho\nu} - \Gamma^\rho_{\mu\nu}\Gamma^\sigma_{\rho\lambda}) V_\sigma$$

changing $\tau \to \sigma$. Hence

$$(V_{\mu;\nu})_{;\lambda} - (V_{\mu;\lambda})_{;\nu} = R^\sigma_{\mu\nu\lambda} V_\sigma = R^\rho_{\mu\nu\lambda} V_\rho.$$

11.2 Show that

$$[\nabla_\mu, \nabla_\nu]V^\rho = R^\rho_{\lambda\mu\nu} V^\lambda.$$

Compare it with electromagnetic case

$$\nabla_\mu = \partial_\mu + ieF_{\mu\nu},$$

$$[\nabla_\mu, \nabla_\nu]V^\rho = ieF_{\mu\nu} V^\rho,$$

implies

$$[\nabla_\mu, \nabla_\nu] = ieF_{\mu\nu}.$$

[$F^{\mu\nu}$: electromagnetic field tensor. Here it is possible to factor out V^ρ in contrast to the gravitational field contained in the curvature tensor because of nonlinearity of general theory of relativity.]

Solution

$$(\nabla_\mu \nabla_\nu - \nabla_\nu \nabla_\mu) V^\rho = \nabla_\mu (\partial_\nu V^\rho + \Gamma^\rho_{\nu\lambda} V^\lambda) - \nabla_\nu (\partial_\mu V^\rho + \Gamma^\rho_{\mu\lambda} V^\lambda)$$

$$= \partial_\mu (\partial_\nu V^\rho + \Gamma^\rho_{\nu\lambda} V^\lambda) + \Gamma^\rho_{\mu\sigma} (\partial_\nu V^\sigma + \Gamma^\sigma_{\nu\lambda} V^\lambda)$$
$$-\Gamma^\sigma_{\mu\nu} (\partial_\sigma V^\rho + \Gamma^\rho_{\sigma\lambda} V^\lambda) - \mu \leftrightarrow \nu$$

$$= [\partial_\mu (\partial_\nu V^\rho + \Gamma^\rho_{\nu\lambda} V^\lambda) + \Gamma^\rho_{\mu\sigma} (\partial_\nu V^\sigma + \Gamma^\sigma_{\nu\lambda} V^\lambda)$$
$$-\Gamma^\sigma_{\mu\nu} (\partial_\sigma V^\rho + \Gamma^\rho_{\sigma\lambda} V^\lambda)] - [\partial_\nu (\partial_\mu V^\rho + \Gamma^\rho_{\mu\lambda} V^\lambda)$$
$$+\Gamma^\rho_{\nu\sigma} (\partial_\mu V^\sigma + \Gamma^\sigma_{\mu\lambda} V^\lambda) - \Gamma^\sigma_{\nu\mu} (\partial_\sigma V^\rho + \Gamma^\rho_{\sigma\lambda} V^\lambda)]$$

$$= \partial_\mu (\Gamma^\rho_{\nu\lambda} V^\lambda) + \Gamma^\rho_{\mu\sigma} (\partial_\nu V^\sigma + \Gamma^\sigma_{\nu\lambda} V^\lambda)$$
$$-\partial_\nu (\Gamma^\rho_{\mu\lambda} V^\lambda) - \Gamma^\rho_{\nu\sigma} (\partial_\mu V^\sigma + \Gamma^\sigma_{\mu\lambda} V^\lambda) - [\Gamma^\sigma_{\mu\nu} - \Gamma^\sigma_{\nu\mu}] \nabla_\sigma V^\rho$$

$$= \partial_\mu (\Gamma^\rho_{\nu\lambda}) V^\lambda + \Gamma^\rho_{\nu\lambda} \partial_\mu V^\lambda + \Gamma^\rho_{\mu\sigma} \partial_\nu V^\sigma + \Gamma^\rho_{\mu\sigma} \Gamma^\sigma_{\nu\lambda} V^\lambda$$
$$-(\partial_\nu \Gamma^\rho_{\mu\lambda}) V^\lambda - \Gamma^\rho_{\mu\lambda} \partial_\nu V^\lambda - \Gamma^\rho_{\nu\sigma} \partial_\mu V^\sigma - \Gamma^\rho_{\nu\sigma} \Gamma^\sigma_{\mu\lambda} V^\lambda$$

$$= [\partial_\mu \Gamma^\rho_{\nu\lambda} - \partial_\nu \Gamma^\rho_{\mu\lambda} + \Gamma^\rho_{\mu\sigma} \Gamma^\sigma_{\nu\lambda} - \Gamma^\rho_{\nu\sigma} \Gamma^\sigma_{\mu\lambda}] V^\lambda = R^\rho_{\lambda\mu\nu} V^\lambda$$

since $\Gamma^\sigma_{\mu\nu} = \Gamma^\sigma_{\nu\mu}$.

11.3 Using Eq. (11.13) of the text derive explicitly the symmetry properties (A), (B) and (C) of curvature tensor $R_{\tau\mu\nu\lambda}$.

11.4 A perfect fluid, in the rest frame of the fluid, i.e., in co-moving frame with velocity **v**, is completely specified by

$$T^{00} = \rho c^2,$$

$$T^{i0} = 0 = T^{0i},$$

$$T^{ij} = p\delta_{ij}.$$

In Minkowski space:

$$T^{\mu\nu} = (\rho c^2 + p) \frac{U^\mu U^\nu}{c^2} - \eta^{\mu\nu} p,$$

$$U^\mu = \frac{dx^\mu}{d\tau} = \gamma \frac{dx^\mu}{dt} = \gamma v^\mu,$$

$$U^0 = \gamma c,$$

$$U^i = \gamma v^i,$$

$$T^{00} = \gamma^2 (\rho c^2 + p).$$

In general coordinate system

$$T^{\mu\nu} = (\rho c^2 + p) \frac{U^\mu U^\nu}{c^2} - g^{\mu\nu} p.$$

Using a perfect fluid source for energy momentum tensor and noting that in the Newtonian limit, p can be put equal to zero, so that

$$T^{\mu\nu} = \rho U^\mu U^\nu,$$

show that in weak gravity limit for the slowly moving particles, the Ricci tensor

$$R_{00} = \frac{1}{2}\kappa\rho c^2.$$

and $\vec{\nabla}^2 \epsilon_{00} = \frac{1}{2}\kappa\rho c^2$.

Solution: In the weak field limit

$$g_{\mu\nu} = \eta_{\mu\nu} + \epsilon_{\mu\nu},$$

$$R_{00} = R^{00} \approx \kappa(T^{00} - \frac{1}{2}g^{00}T_0^0)$$

$$\approx \frac{1}{2}\kappa\rho c^2, \tag{11.65}$$

$$\Gamma_{\mu\nu}^\lambda \approx \frac{1}{2}\eta^{\lambda\sigma}(\partial_\nu\epsilon_{\mu\sigma} + \partial_\mu\epsilon_{\nu\sigma} - \partial_\sigma\epsilon_{\mu\nu}).$$

Thus in this limit

$$R_{\tau\mu\nu\lambda} \approx \frac{1}{2}(\partial_\lambda\partial_\tau\epsilon_{\mu\nu} - \partial_\lambda\partial_\mu\epsilon_{\nu\tau} + \partial_\nu\partial_\mu\epsilon_{\lambda\tau} - \partial_\nu\partial_\tau\epsilon_{\mu\lambda}).$$

Now

$$R_{\mu\nu} = R_{\mu\lambda\nu}^\lambda = g^{\tau\lambda}R_{\tau\mu\lambda\nu} = -g^{\tau\lambda}R_{\tau\mu\nu\lambda}$$

$$= -\frac{1}{2}\eta^{\tau\lambda}(\partial_\lambda\partial_\tau\epsilon_{\mu\nu} - \partial_\lambda\partial_\mu\epsilon_{\nu\tau} + \partial_\nu\partial_\mu\epsilon_{\lambda\tau} - \partial_\nu\partial_\tau\epsilon_{\mu\lambda}).$$

Therefore

$$R_{00} = R_{0\lambda0}^\lambda = R_{000}^0 + R_{0i0}^i$$

$$= R_{0i0}^i.$$

$$R_{0i0}^i = -g^{\tau i}R_{\tau00i}$$

$$= -\eta^{ji}R_{j00i}$$

$$= -\eta^{ji}\frac{1}{2}(\partial_i\partial_j\epsilon_{00} - \partial_i\partial_0\epsilon_{0j} + \partial_0\partial_0\epsilon_{ij} - \partial_0\partial_j\epsilon_{0i})$$

$$= \frac{1}{2}(-\partial_i\partial^i\epsilon_{00} + \partial^j\partial_0\epsilon_{0j} - \partial_0\partial_0\epsilon_i^i + \partial_0\partial^i\epsilon_{0i})$$

$$= -\frac{1}{2}\partial_i\partial^i\epsilon_{00} = \frac{1}{2}\nabla^2\epsilon_0,$$

since for static field, ϵs are independent of t.

Thus

$$R_{00} = \frac{1}{2}\nabla^2 \epsilon_{00}. \tag{11.66}$$

Hence from Eqs. (11.65) and (11.66)

$$R_{00} = \frac{1}{2}\vec{\nabla}^2 \epsilon_{00} = \frac{1}{2}\kappa\rho c^2.$$

Comparing it with Poisson equation in Newtonian gravity:

$$\nabla^2 \Phi = 4\pi G\rho, \tag{11.67}$$

$$\Phi = \frac{c^2}{2}\epsilon_{00}, \tag{11.68}$$

$$\kappa = \frac{8\pi G}{c^4}. \tag{11.69}$$

11.5 For the metric

$$ds^2 = c^2 dt^2 - a^2(t)[dx^{1^2} + dx^{2^2} + dx^{3^2}],$$

(i) calculate the Ricci tensor and Ricci scalar.
(ii) Find the geodesic.

Chapter 12

THE SCHWARZSCHILD, FRIEDMANN ROBERTSON WALKER METRIC

12.1   Introduction

Einstein's field equations, being nonlinear are difficult to solve exactly. In general relativity the metric is given by

$$ds^2 = g_{\mu\nu}dx^\mu dx^\nu$$

$$= g_{00}c^2 dt^2 + g_{ij}dx^i dx^j + g_{oj}dx^0 dx^j. \qquad (12.1)$$

We will consider only the metric for which the off-diagonal terms can be eliminated by suitable transformations so that the metric tensor $g_{\mu\nu}$ is diagonal:

$$g_{\mu\nu} = 0, \ \mu \neq \nu$$

$$g_{\mu\nu} = h_\mu \delta_{\mu\nu}, \ g^{\mu\nu} = \frac{1}{h_\mu}\delta^{\mu\nu}. \qquad (12.2)$$

In this case, we have the orthogonal coordinate system. Thus the Christoffel symbols

$$\Gamma^\lambda_{\mu\nu} = \frac{1}{2}g^{\lambda\sigma}\left[\partial_\nu g_{\mu\sigma} + \partial_\mu g_{\nu\sigma} - \partial_\sigma g_{\mu\nu}\right] \qquad (12.3)$$

are given by

$$\Gamma^\lambda_{\mu\nu} = 0, \ \lambda \neq \mu \neq \nu,$$

$$\Gamma^\lambda_{\mu\lambda} = \frac{1}{2}g^{\lambda\lambda}\left[\frac{\partial g_{\mu\lambda}}{\partial x^\lambda} + \frac{\partial g_{\lambda\lambda}}{\partial x^\mu} - \frac{\partial g_{\mu\lambda}}{\partial x^\lambda}\right]$$

$$= \frac{1}{2}\frac{1}{h_\lambda}\frac{\partial h_\lambda}{\partial x^\mu} = \Gamma^\lambda_{\lambda\mu}, \qquad (12.4)$$

$$\Gamma^\lambda_{\lambda\lambda} = \frac{1}{2h_\lambda}\frac{\partial h_\lambda}{\partial x^\lambda},$$

$$\Gamma^\lambda_{\mu\mu} = -\frac{1}{2h_\lambda}\frac{\partial h_\mu}{\partial x^\lambda}.$$

12.2   The Schwarzschild Metric

It was noted by Schwarzschild that in a special case, of the static spherically symmetric field generated by a spherical mass M at rest, Einstein's
equations can be solved without much difficulty. The solution describes the
relativistic generalization of the Newton's solution namely, gravitational potential $\Phi = -\frac{GM}{r}$, M being the gravitational mass of the spherical source
distribution and r is the radial coordinate from the centre of the mass
distribution.

The most general form of ds^2 compatible with spherical symmetry is

$$ds^2 = A(r,t)c^2 dt^2 - C(r,t)(d\theta^2 + \sin^2\theta d\phi^2) - D(r,t)cdtdr$$
$$-E(r,t)dr^2 \tag{12.5}$$

where A, C, D and E are in general functions of r, t. The center of symmetry
is the point $r = 0$.

We now try to eliminate the off-diagonal term by making the transformation

$$dt' = dt - \frac{1}{2}\frac{D}{AC}dr. \tag{12.6}$$

Then ds^2 takes the form

$$ds^2 = A(r,t')c^2 dt'^2 - B(r,t')dr^2 - C(r,t')(d\theta^2 + \sin^2\theta d\phi^2), \tag{12.7}$$

where

$$B = E + \frac{D^2}{4AC^2}. \tag{12.8}$$

Removing the prime and making use of the static conditions so that A, B
and C are functions of r alone, we can write

$$ds^2 = A(r)c^2 dt^2 - B(r)dr^2 - C(r)(d\theta^2 + \sin^2\theta d\phi^2). \tag{12.9}$$

We now introduce a new radial coordinate R, the Schwarzschild radial coordinate:

$$\sqrt{C(r)} = R. \tag{12.10}$$

Then

$$Bdr^2 = 4BC(\frac{dC}{dr})^{-2}dR^2 \equiv \overline{B}dR^2. \tag{12.11}$$

By means of Eq. (12.10) we can express $\overline{B}$ as a function of radial coordinate
R. Thus

$$ds^2 = Ac^2 dt^2 - \overline{B}dr^2 - r^2(d\theta^2 + \sin^2\theta d\phi^2) \tag{12.12}$$

where instead of R, r is now the Schwarzschild radial coordinate. This involves two unknown functions of r. In order to ensure the signature of $g_{\mu\nu}$ and the sign of $\det g_{\mu\nu} = g$, it is usual to write $A(r)$ as an intrinsically positive function $e^{2\lambda(r)}$ and $\overline{B}(r)$ as $e^{2\nu(r)}$ so that the line element is written as

$$ds^2 = e^{2\lambda(r)}c^2 dt^2 - e^{2\nu(r)}dr^2 - r^2(d\theta^2 + \sin^2\theta d\phi^2). \tag{12.13}$$

This form of the line element meets clearly the demands of time-independence and radial symmetry and will be used to obtain the Schwarzschild solution. The Schwarzschild radial coordinate has a special meaning as the area of surface of sphere $r =$ constant is given by $4\pi r^2$ as in flat space, which allows us to determine the value of r.

From Eq. (12.13)

$$g_{00} = h_0 = e^{2\lambda(r)}, \quad g_{11} = h_1 = -e^{2\nu(r)},$$
$$g_{22} = h_2 = -r^2, g_{33} = h_3 = -r^2\sin^2\theta \tag{12.14}$$

since all hs are independent of time, the only non-zero components of $\Gamma^\lambda_{\mu\nu}$ (cf. Eq. (12.4)) are

$$\Gamma^0_{10} = \frac{1}{2h_0}\frac{\partial h_0}{\partial r} = \lambda'$$

$$\Gamma^1_{00} = -\frac{1}{2h_1}\frac{\partial h_0}{\partial r} = \lambda' e^{2(\lambda-\nu)}$$

$$\Gamma^1_{11} = \frac{1}{2h_1}\frac{\partial h_1}{\partial r} = \nu'$$

$$\Gamma^1_{22} = -\frac{1}{2h_1}\frac{\partial h_2}{\partial r} = -re^{-2\nu} \tag{12.15}$$

$$\Gamma^1_{33} = -\frac{1}{2h_1}\frac{\partial h_3}{\partial r} = -r\sin^2\theta e^{-2\nu}$$

$$\Gamma^2_{12} = \frac{1}{2h_2}\frac{\partial h_2}{\partial r} = \frac{1}{r}$$

$$\Gamma^3_{13} = \frac{1}{2h_3}\frac{\partial h_3}{\partial r} = \frac{1}{r}$$

$$\Gamma^3_{23} = \frac{1}{2h_3}\frac{\partial h_3}{\partial \theta} = \cot\theta.$$

These expressions are then used to calculate Ricci tensor $R_{\mu\nu}$ given by

$$R_{\mu\nu} = \partial_\alpha\Gamma^\alpha_{\mu\nu} - \partial_\nu\Gamma^\alpha_{\mu\alpha} + \Gamma^\alpha_{\mu\nu}\Gamma^\beta_{\alpha\beta} - \Gamma^\alpha_{\mu\beta}\Gamma^\beta_{\nu\alpha}. \tag{12.16}$$

Thus

$$R_{00} = \partial_\alpha \Gamma^\alpha_{00} + \Gamma^\alpha_{00}\Gamma^\beta_{\alpha\beta} - \Gamma^\alpha_{0\beta}\Gamma^\beta_{0\alpha}$$

$$= \partial_1 \Gamma^1_{00} + \Gamma^1_{00}(\Gamma^0_{10} + \Gamma^1_{11} + \Gamma^2_{12} + \Gamma^3_{13}) - \Gamma^0_{01}\Gamma^1_{00} - \Gamma^1_{00}\Gamma^0_{01}$$

$$= \lambda'' - \lambda'\nu' + \lambda'^2 + \frac{2\lambda'}{r}e^{2(\lambda-\nu)}. \tag{12.17}$$

Similarly

$$R_{11} = -\partial_1 \Gamma^\alpha_{1\alpha} + \partial_\alpha \Gamma^\alpha_{11} + \Gamma^\alpha_{11}\Gamma^\beta_{\alpha\beta} - \Gamma^\alpha_{1\beta}\Gamma^\beta_{1\alpha}$$

$$= -\lambda'' + \nu'\lambda' - \lambda'^2 + \frac{2\nu'}{r}e^{2(\lambda-\nu)} \tag{12.18}$$

$$R_{22} = -\partial_2 \Gamma^\alpha_{2\alpha} + \partial_\alpha \Gamma^\alpha_{22} + \Gamma^\alpha_{22}\Gamma^\beta_{\alpha\beta} - \Gamma^\alpha_{2\beta}\Gamma^\beta_{2\alpha}$$

$$= -\partial_2(\cot\theta) + \partial_1 \Gamma^1_{22} + \Gamma^1_{22}\Gamma^\beta_{1\beta} - \Gamma^1_{22}\Gamma^2_{21}$$

$$-\Gamma^2_{21}\Gamma^1_{22} - \Gamma^3_{23}\Gamma^3_{23}$$

$$= 1 - e^{-2\nu}(1 - r\nu' + r\lambda') \tag{12.19}$$

$$R_{33} = \sin^2\theta R_{22}. \tag{12.20}$$

Now in empty space Einstein's equations give

$$R_{\mu\nu} = 0 \tag{12.21}$$

so that we have from Eqs. (12.17)–(12.19)

$$\lambda'' - \lambda'\nu' + \lambda'^2 + \frac{2\lambda'}{r}e^{2(\lambda-2)} = 0 \tag{12.22}$$

$$-\lambda'' + \lambda'\nu' - \lambda'^2 + \frac{2\nu'}{r}e^{2(\lambda-2)} = 0 \tag{12.23}$$

$$1 - e^{-2\nu}(1 - r\nu' + r\lambda') = 0. \tag{12.24}$$

Adding Eqs. (12.22) and (12.23), we obtain

$$\lambda' + \nu' = 0. \tag{12.25}$$

Its integration gives

$$\lambda + \nu = const.r. \tag{12.26}$$

But as $r \longrightarrow \infty$, we have flat space so that from Eq. (12.13)

$$\lambda = 0, \ \nu = 0 \tag{12.27}$$

in this limit and hence constant in Eq. (12.26) must be zero. Thus

$$\lambda(r) = -\nu(r). \tag{12.28}$$

Further Eqs. (12.24) and (12.25) give

$$\frac{d}{dr}[re^{-2\nu}] = 1. \tag{12.29}$$

Integration gives

$$re^{-2\nu} = r + b, \tag{12.30}$$

where b is a constant of integration. Thus $[\nu = -\lambda]$

$$g_{00} = e^{2\lambda} = 1 + \frac{b}{r}. \tag{12.31}$$

Now as already seen in Chapter 10 that in the Newtonian limit of field equations,

$$g_{00} = 1 + \frac{\Phi}{c^2}, \tag{12.32}$$

where Φ is the Newtonian potential

$$\Phi = \frac{-GM}{r} \tag{12.33}$$

due to a body of mass M. Thus in this limit

$$ds^2 = c^2\left(1 + \frac{\Phi}{c^2}\right)dt^2 - \left(1 + \frac{\Phi}{c^2}\right)^{-1}dr^2 - r^2\left(d\theta^2 + \sin^2\theta d\phi^2\right). \tag{12.34}$$

However, Eqs. (12.32), (12.33) leads to the identification of the constant b in Eq. (12.31):

$$b = \frac{-2GM}{c^2} = -r_s, \tag{12.35}$$

where r_s is known as the Schwarzschild radius. Thus

$$g_{00} = e^{2\lambda} = 1 - \frac{r_s}{r}, \quad g_{11} = g_{rr} = -e^{2\nu} = -e^{-2\lambda} = -(1 - \frac{r_s}{r})^{-1}. \tag{12.36}$$

Hence the Schwarzschild metric is given by

$$ds^2 = (1 - \frac{r_s}{r})c^2dt^2 - (1 - \frac{r_s}{r})^{-1}dr^2 - r^2(d\theta^2 + \sin^2\theta d\phi^2). \tag{12.37}$$

This solution tells us that "the empty spacetime outside a spherically symmetric distribution of matter is describable by a static metric". This is known as Birkhoff's theorem. The following remarks are relevant:

(i) The line element (12.37) becomes singular at $r = r_s$, since $g_{00} = 0$ and g_{11} approaches $-\infty$. This is known as Schwarzschild singularity.

(ii) For known spherical bodies r_s lies within the body, where Schwarzschild solution is not applicable anyway. For example, r_s for the sun is $\frac{2GM_\odot^2}{c^2} = 2.95$ km much smaller than the radius of the solar surface 6.95×10^5 km.

At the surface the Schwarzschild geometry joins a different geometry inside a star. As long as one is outside of static stars, one does not have to worry about radii $r = 0$ and $r = r_s$. Nevertheless one has to worry about these radii when gravitational collapse occurs. Interesting things do occur at the boundary $r = r_s$. For example, matter and energy may fall into the region but nothing can come out of it, a region of so-called black hole.

12.3 Friedmannn-Robertson-Walker (FRW) Metric

Before we discuss the FRW metric, it is convenient to discuss some geometrical concepts for better understanding of this metric.

In [4] Euclidean space, introduce the spherical polar coordinates

$$
\begin{aligned}
x &= r \sin \chi \sin \theta \cos \phi \\
y &= r \sin \chi \sin \theta \sin \phi \\
z &= r \sin \chi \cos \theta \\
u &= r \cos \chi
\end{aligned}
\tag{12.38}
$$

The metric in [4] space

$$
dS^2 = dr^2 + r^2 d\chi^2 + r^2 \sin^2 \chi \left(d\theta^2 + \sin^2 \theta d\phi^2 \right).
\tag{12.39}
$$

(1) Consider [3] sphere embedded in [4] Euclidean space

$$
\begin{aligned}
x &= R \sin \chi \sin \theta \cos \phi \\
y &= R \sin \chi \sin \theta \sin \phi \\
z &= R \sin \chi \cos \theta \\
u &= R \cos \chi
\end{aligned}
\tag{12.40}
$$

$$
x^2 + y^2 + z^2 + u^2 = R^2
$$

R is the radius of the sphere. Then

$$
dS^2 = R^2 \left[d\chi^2 + \sin^2 \chi \left(d\theta^2 + \sin^2 \theta d\phi^2 \right) \right]
\tag{12.41}
$$

a closed curve.

(2) Corresponding to the sphere embedded in [4] space, consider [3] hyperboloid embedded in [4]

$$
\begin{aligned}
x &= R \sinh \chi \sin \theta \cos \phi \\
y &= R \sinh \chi \sin \theta \sin \phi \\
z &= R \sinh \chi \cos \theta \\
u &= R \cosh \chi
\end{aligned}
\tag{12.42}
$$

$$
u^2 - x^2 - y^2 - z^2 = R^2.
$$

Hyperboloid in [4] Minkowski space

$$dS^2 = R^2 \left[d\chi^2 + \sinh^2 \chi \left(d\theta^2 + \sin^2 \theta d\phi^2 \right) \right] \tag{12.43}$$

an open curve.

(3) For flat curve

$$dS^2 = R^2 \left[d\chi^2 + \chi^2 \left(d\theta^2 + \sin^2 \theta d\phi^2 \right) \right]. \tag{12.44}$$

The FRW metric:

I): Closed Universe:

$$\begin{aligned} ds^2 &= c^2 dt^2 - dS^2 \\ &= c^2 dt^2 - R^2 \left(t \right) \left[d\chi^2 + \sin^2 \chi \left(d\theta^2 + \sin^2 \theta d\phi^2 \right) \right] \end{aligned} \tag{12.45}$$

II): Open Universe:

$$ds^2 = c^2 dt^2 - R^2 \left(t \right) \left[d\chi^2 + \sinh^2 \chi \left(d\theta^2 + \sin^2 \theta d\phi^2 \right) \right] \tag{12.46}$$

III): Flat Universe:

$$ds^2 = c^2 dt^2 - R^2 \left(t \right) \left[d\chi^2 + \chi^2 \left(d\theta^2 + \sin^2 \theta d\phi^2 \right) \right] \tag{12.47}$$

where $R \left(t \right)$ is a scale factor.

Define

$$(i): \ r' = \frac{r}{R} = \sin \chi, \ d\chi^2 = \frac{dr'^2}{\cos^2 \chi} = \frac{dr'^2}{1 - r'^2}$$

$$(ii): \ r' = \sinh \chi, \ d\chi^2 = \frac{dr'^2}{1 + r'^2}$$

$$(iii): \ r' = \chi, \ d\chi = dr'.$$

Change $r' \to r$, the FRW metric is given by

$$ds^2 = c^2 dt^2 - R^2 \left(t \right) \left[\frac{dr^2}{1 - kr^2} + r^2 \left(d\theta^2 + \sin^2 \theta d\phi^2 \right) \right] \tag{12.48}$$

where

$$k = \begin{cases} 1 & : \text{Closed Universe} \\ 0 & : \text{Flat Universe} \\ -1 & : \text{Open Universe.} \end{cases}$$

12.4   Cosmology

12.4.1   *Cosmological Principle*

On a sufficiently large scale, universe is homogeneous and isotropic. A coordinate system in which matter is at rest at any moment is called co-moving coordinate system. Any observer in this coordinate system is called co-moving observer. Any co-moving observer will see around himself a uniform and isotropic universe.

Cosmological principle implies existence of a universal time, since all observers see the same sequence of events with which to synchronize their clocks. In particular, they can all start their clocks with Big Bang.

A homogeneous and isotropic Universe is described by the FRW metric

$$ds^2 = c^2 dt^2 - R^2(t) \left[\frac{dr^2}{1 - kr^2} + r^2 \left(d\theta^2 + \sin^2 \theta d\phi^2 \right) \right]$$

r, θ, ϕ are co-moving coordinates and scale factor $R(t)$ describes the expansion.

Now r being a co-moving coordinate is a label for a particular galaxy and does not change as the universe expands. Also for an isotropic universe θ and ϕ remain fixed for a particular galaxy. This means in a give direction (viz the same value of θ and ϕ), the physical distance between two points (or galaxies) say, at $r = 0$ and $r = r'$ is given by

$$\ell = R(t) r'$$

and the velocity of expansion is given by

$$v = \frac{d\ell}{dt} = \dot{R}(t) r'$$

$$= \frac{\dot{R}(t)}{R(t)} R(t) r' = H(t) \ell \equiv H\ell$$

where

$$H = \frac{\dot{R}(t)}{R(t)}$$

is the Hubble parameter.

Furthermore for a light signal $ds^2 = 0$, thus for fixed θ and ϕ

$$c^2 dt^2 = R^2(t) \left[\frac{dr^2}{1 - kr^2} \right]$$

$$cdt = R(t) \left[\frac{dr}{\sqrt{1 - kr^2}} \right]. \tag{12.49}$$

Thus a light signal leaves distant galaxy travelling along $-r$ direction, at time t_e. θ_e, ϕ_e, will reach the place where it is observed in time

$$c \int_{t_e}^{t_o} \frac{dt}{R(t)} = f(r_e) \tag{12.50}$$

$$f(r_e) = \int_0^{r_e} \frac{dr}{\sqrt{1 - kr^2}} = \begin{cases} \sin^{-1} r_e, & k = 1 \\ r_e, & k = 0 \\ \sinh^{-1} r_e, & k = -1. \end{cases}$$

If a series of light pulses in the distant galaxies are emitted at equal short time interval δt_e, $\nu_e = \frac{1}{\delta(t_e)}$ and

$$c \int_{t_e + \delta t_e}^{t_o + \delta t_o} \frac{dt}{R(t)} = f(r_e). \tag{12.51}$$

Thus

$$\frac{\delta t_o}{R(t_o)} = \frac{\delta t_e}{R(t)}, \quad \frac{\delta t_o}{\delta t_e} = \frac{R(t_o)}{R(t)}.$$

Hence

$$\frac{\nu_e}{\nu_o} = \frac{R(t_o)}{R(t)} \tag{12.52}$$

or

$$\frac{\lambda_o}{\lambda_e} = \frac{R(t_o)}{R(t)}.$$

Define the redshift

$$z = \frac{\Delta \lambda}{\lambda_e} = \frac{\lambda_o - \lambda_e}{\lambda_e} = \frac{\lambda_o}{\lambda_e} - 1$$

or

$$1 + z = \frac{\lambda_o}{\lambda_e} = \frac{R(t_o)}{R(t)} \equiv \frac{R_o}{R(t)}. \tag{12.53}$$

Thus

$$H(t) = H(z) = \frac{\dot{R}(t)}{R(t)} = \frac{\dot{R}(t)}{R_o} \frac{R_o}{R(t)} = (1 + z) \frac{\dot{R}(t)}{R_o}.$$

Therefore

$$\frac{\dot{R}(t)}{R_o} = \frac{H(z)}{1 + z}. \tag{12.54}$$

Thus the first consequence of cosmological principle is redshift of spectral lines for the light received from distant galaxies. The above equation shows that the redshift directly indicates the relative linear size of the universe when the photon was emitted. The redshift is experimentally observed and it clearly shows that universe is expanding. Highest redshift so far observed $z = 6.96$, so that Lyman alpha line appears in the red part of spectrum around 7200 A^o. This implies

$$\frac{R(t_o)}{R(t_e)} = 1 + z = 7.96.$$

The essential features of the standard model of cosmology that the Universe "started" in a hot and dense phase and has been undergoing expansion since it is based on two directly observed facts:

a): The distant cosmological objects were found to be moving away from observers with velocities proportional to the their distance, i.e., $H\ell$.

b): The Universe is filled with gas of photons with temperature $T_\circ \simeq$ 2.7K called the Cosmic Microwave Background (CMB) radiation. This is supposed to be relic of the early Universe. The observed isotropy of the CMB radiation $(\Delta T/\Gamma \sim 10^{-5})$ provide the strongest direct support of the cosmological principle.

12.4.2   *Standard Model of Cosmology*

On sufficiently large scale, Universe is homogeneous and isotropic, described by the FRW metric

$$ds^2 = c^2 dt^2 - R^2(t)\left[\frac{dr^2}{1-kr^2} + r^2\left(d\theta^2 + \sin^2\theta d\phi^2\right)\right].$$

For this metric

$$g_{00} = h_0 = 1, \; g_{11} = h_1 = -\frac{R^2(t)}{1-kr^2},$$

$$g_{22} = h_2 = -R^2(t)r^2, \; g_{33} = h_3 = -R^2(t)r^2\sin^2\theta. \qquad (12.55)$$

Einstein's equations

$$R_{\mu\nu} = \frac{8\pi G}{c^4}\left(T_{\mu\nu} - \frac{1}{2}g_{\mu\nu}T^\lambda_\lambda\right) - g_{\mu\nu}\Lambda \qquad (12.56)$$

$$R_{\mu\nu} = \partial_\lambda\Gamma^\lambda_{\mu\nu} - \partial_\nu\Gamma^\lambda_{\mu\lambda} + \Gamma^\rho_{\mu\nu}\Gamma^\lambda_{\rho\lambda} - \Gamma^\rho_{\mu\lambda}\Gamma^\lambda_{\rho\nu}. \qquad (12.57)$$

From Eq. (12.4), since $h_0 = 1$

$$\Gamma^0_{00} = 0, \; \Gamma^0_{i0} = \Gamma^0_{0i} = 0 \qquad (12.58)$$

$$\Gamma^0_{ii} = -\frac{1}{2} \frac{\partial h_i}{\partial x^0}$$

$$\Gamma^0_{11} = \frac{1}{c} \left(\frac{R\dot{R}}{1 - kr^2} \right), \quad \Gamma^0_{22} = \frac{1}{c} \frac{R\dot{R}}{r^2}, \quad \Gamma^0_{33} = \frac{R\dot{R}}{cr^2 \sin\theta}. \tag{12.59}$$

For $\lambda = i$

$$\Gamma^i_{0i} = \frac{1}{2h_i} \frac{\partial h_i}{\partial x^0}, \quad \Gamma^i_{ji} = \Gamma^i_{ij} = \frac{1}{2h_i} \frac{\partial h_i}{\partial x^j}$$

$$\Gamma^i_{ii} = \frac{1}{2h_i} \frac{\partial h_i}{\partial x^i}, \quad \Gamma^i_{jj} = -\frac{1}{2h_j} \frac{\partial h_j}{\partial x^i}. \tag{12.60}$$

For the Ricci tensors R_{00}, R_{11}, the non-zero Γs are

$$\Gamma^0_{11} = \frac{1}{c} \left(\frac{R\dot{R}}{1 - kr^2} \right) \tag{12.61}$$

$$\Gamma^1_{01} = \frac{1}{h_1} \frac{\partial h_1}{\partial x^0} = \frac{1}{c} \frac{\dot{R}}{R} = \Gamma^2_{02} = \Gamma^3_{03} \tag{12.62}$$

$$\Gamma^1_{11} = \frac{1}{2h_1} \frac{\partial h_1}{\partial r} = \frac{kr}{1 - kr^2} \tag{12.63}$$

$$\Gamma^2_{12} = \frac{1}{2h_2} \frac{\partial h_2}{\partial r} = \frac{1}{r}, \quad \Gamma^3_{13} = \frac{1}{2h_3} \frac{\partial h_3}{\partial r} = \frac{1}{r}. \tag{12.64}$$

Thus, using Eqs. (12.62)–(12.64)

$$R_{00} = \partial_\lambda \Gamma^\lambda_{00} - \partial_0 \Gamma^\lambda_{0\lambda} + \Gamma^\rho_{00}\Gamma^\lambda_{\rho\lambda} - \Gamma^\rho_{0\lambda}\Gamma^\lambda_{\rho 0}$$

$$= -\partial_0 \Gamma^i_{0i} - \Gamma^0_{0i}\Gamma^i_{0i} = -\frac{3}{c^2} \left[\frac{\ddot{R}}{R} - \frac{\dot{R}^2}{R^2} \right] - \frac{3}{c^2} \left(\frac{\dot{R}}{R} \right)^2$$

$$= -\frac{3}{c^2} \frac{\ddot{R}}{R}, \tag{12.65}$$

$$R_{11} = \partial_\lambda \Gamma^\lambda_{11} - \partial_1 \Gamma^\lambda_{1\lambda} + \Gamma^\rho_{11}\Gamma^\lambda_{\rho\lambda} - \Gamma^\rho_{1\lambda}\Gamma^\lambda_{1\rho}$$

$$= \partial_0 \Gamma^0_{11} - \partial_r \left(\Gamma^2_{12} + \Gamma^3_{13} \right) + \Gamma^0_{11} \left(\Gamma^1_{01} + \Gamma^2_{02} + \Gamma^3_{03} \right)$$

$$+ \Gamma^1_{11} \left(\Gamma^1_{11} + \Gamma^2_{12} + \Gamma^3_{13} \right) - 2\Gamma^0_{11}\Gamma^1_{01}$$

$$- \left(\Gamma^1_{11}\Gamma^1_{11} + \Gamma^2_{12}\Gamma^2_{12} + \Gamma^3_{13}\Gamma^3_{13} \right)$$

$$= \frac{R\ddot{R} + 2\dot{R}^2 + 2kc^2}{c^2 \left(1 - kr^2 \right)}. \tag{12.66}$$

12.4.3 Friedmann-Lemaitre Equations

It is customary to take for the energy momentum tensor $T_{\mu\nu}$ on the R.H.S. of Eq. (12.56) as that of a perfect fluid

$$T_{\mu\nu} = \left(\rho c^2 + p\right)\frac{U_\mu U_\nu}{c^2} - g_{\mu\nu}p. \tag{12.67}$$

In a co-moving frame, $U_i = 0$, $U_0 = c$, $g_{00} = 1$. Thus

$$T_{00} = \rho c^2, \ \Gamma_{0i} = \Gamma_{i0} = 0$$

$$T_{ij} = -g_{ij}p = -ph_i\delta_{ij}. \tag{12.68}$$

Now

$$T_\nu^\mu = g^{\lambda\mu}T_{\lambda\nu}. \tag{12.69}$$

Thus

$$T_0^0 = T_{00} = \rho c^2, \ T_i^0 = 0 = T_0^i$$

$$T_j^i = g^{ki}T_{kj} = g^{ki}\left(-g_{kj}p\right) = -\delta_j^i p$$

$$T_\lambda^\lambda = \left(\rho c^2, -p, -p, -p\right). \tag{12.70}$$

Hence from Eq. (12.56) using Eqs. (12.68)–(12.70)

$$R_{00} = \frac{8\pi G}{c^4}\left(T_{00} - \frac{1}{2}T_\lambda^\lambda\right) - g_{00}\Lambda$$

$$= \frac{4\pi G}{c^4}\left(\rho c^2 + 3p\right) - \Lambda, \tag{12.71}$$

$$R_{11} = \frac{8\pi G}{c^4}\left(T_{11} - \frac{1}{2}g_{11}T_\lambda^\lambda\right) - g_{11}\Lambda$$

$$= g_{11}\left[\frac{4\pi G}{c^4}\left(p - \rho c^2\right) - \Lambda\right] \tag{12.72}$$

$$= \frac{R^2}{1 - kr^2}\left[\frac{4\pi G}{c^4}\left(\rho c^2 - p\right) + \Lambda\right].$$

Thus from Eqs. (12.65), (12.66) and Eqs. (12.71), (12.72), we obtain

$$\frac{\ddot{R}}{R} = \frac{1}{3}\Lambda c^2 - \frac{4\pi G}{3c^2}\left(\rho c^2 + 3p\right), \tag{12.73}$$

$$\frac{\ddot{R}}{R} + 2\frac{\dot{R}^2}{R^2} + \frac{2kc}{R^2} = \frac{4\pi G}{c^2}\left(\rho c^2 - p\right) + \Lambda c^2. \tag{12.74}$$

Subtracting Eq. (12.73) from Eq. (12.74), one obtains

$$H^2 = \left(\frac{\dot{R}}{R}\right)^2 = \frac{8\pi G}{3}\rho - \frac{kc^2}{R^2} + \frac{1}{3}\Lambda c^2 \tag{12.75}$$

Eqs. (12.73) and (12.75) are called the Friedmann-Lemaitre equations. $H(t)$ is the Hubble parameter.

Now, the energy conservation gives

$$\nabla_\mu T_0^\mu = 0, \tag{12.76}$$

where

$$\begin{aligned}
\nabla_\mu T_0^\mu &= \partial_\mu T_0^\mu + \Gamma_{\mu\lambda}^\mu T_0^\lambda - \Gamma_{\mu 0}^\lambda T_\lambda^\mu \\
&= \partial_0 T_0^0 + \Gamma_{i0}^i T_0^0 - \Gamma_{i0}^j T_j^i \\
&= \partial_0\left(\rho c^2\right) + \rho c^2 \Gamma_{i0}^i + p\delta_j^i \Gamma_{i0}^j \\
&= c\dot{\rho} + \left(\rho c^2 + p\right)\frac{3\dot{R}}{cR}.
\end{aligned} \tag{12.77}$$

The energy conservation gives

$$\dot{\rho} = -\frac{3}{c^2}\frac{\dot{R}}{R}\left(\rho c^2 + p\right) = -\frac{3H}{c^2}\left(\rho c^2 + p\right). \tag{12.78}$$

The same result follows from the first law of thermodynamics

$$d\left(\rho c^2 V\right) + pdV = 0, \quad V \propto R^3,$$

$$\frac{d}{dt}\left(\rho c^2 R^3\right) + \frac{d}{dt}R^3 = 0.$$

In addition, equation of state is needed which is taken to be that of an ideal gas:

$$p = nk_B T,$$

where n is the particle density k_B is the Boltzmann constant $\left(k_B = 0.86 \times 10^{-10} \text{ MeV.K}^{-1}\right)$. For the non-relativistic gas (NR), $k_B T \ll mnc^2$, i.e. $p \ll \rho c^2$, $\rho = mn$, we say that Universe is matter dominated. For extreme relativistic gas (ER): $p = \frac{1}{3}\rho c^2$, $\rho c^2 = 3nk_B T$ and we say that Universe is radiation dominated. Thus we can write the equation of state

$$p = \omega\rho c^2. \tag{12.79}$$

The three cases

$$\omega = \begin{cases} 0, \text{ non-relativistic gas: matter} \\ 1/3, \text{ relativistic gas: radiation} \\ -1, \text{ negative pressure} \end{cases}$$

12.4.4   *Observational Cosmology*

From Eqs. (12.78) and (12.79)

$$\frac{\dot{\rho}}{\dot{R}} = -3\left(1+\omega\right)\frac{\rho}{R}$$

or

$$\frac{\dot{\rho}}{\rho} = -3\left(1+\omega\right)\frac{\dot{R}}{R} \tag{12.80}$$

$$\ln\rho = \ln R^{-3(1+\omega)}.$$

Thus

$$\rho \sim \frac{1}{R^{3(1+\omega)}} \sim \begin{cases} 1/R^3, & \omega = 0\text{: matter dominated} \\ 1/R^4, & \omega = 1/3\text{: radiation dominated} \\ \text{constant}, & \omega = -1 \end{cases} \tag{12.81}$$

We rewrite Eq. (12.75) in the form

$$kc^2 = R^2\left(t\right)H^2\left(t\right)\left[\frac{8\pi G\rho}{3H^2\left(t\right)} + \frac{\Lambda c^2}{3H^2\left(t\right)} - 1\right]. \tag{12.82}$$

Friedmann's equations in the form given in Eqs. (12.73) and (12.82) are the fundamental equations for observational cosmology.

12.4.4.1   *Cosmological parameters*

(i): Hubble parameter

The Hubble parameter $H\left(t\right)$ or $H\left(z\right)$ and the redshift z are given in Eqs. (12.53) and (12.54). The numerical values of Hubble parameter is given below

$$H = 100 \text{ h km s}^{-1}\text{Mpc}^{-1}$$

$$1\text{pc} = 3.0856 \times 10^{18} \text{ cm} = 3.0856 \times 10^{13} \text{ km} = 3.2615 \text{ light years}$$

$$1\text{Mpc} = 3.0856 \times 10^{19} \text{ km}$$

$$h_0 = 0.710$$

$$G = 6.674 \times 10^{-23} \text{ km}^3/\text{gm}.$$

$$H_0 = 100 h_0 \text{ km s}^{-1}\text{Mpc}^{-1}$$

$$H_0^{-1} \approx 9.78 h_0^{-1} \text{ } Gyr \approx 13.77 \text{ } Gyr$$

(ii): Critical density

$$\rho_c(t) = \frac{3H^2(t)}{8\pi G}, \quad \Omega(t) = \frac{\rho(t)}{\rho_c(t)}$$

$$\rho(t) = \Omega(t)\rho_c(t) \equiv \rho. \tag{12.83}$$

The energy density has several components: $\rho_i(t) = \Omega_i(t)\rho_c(t)$.

(iii): Deceleration parameter

$$q(t) = -\frac{\ddot{R}(t)R(t)}{\dot{R}^2(t)} = -\frac{\ddot{R}(t)R^2(t)}{R(t)\dot{R}^2(t)} = -\frac{\ddot{R}(t)}{R(t)}\frac{1}{H^2(t)}. \tag{12.84}$$

Define the cosmological constant in terms of the vacuum energy

$$\Lambda c^2 \to 8\pi G\rho_V. \tag{12.85}$$

In terms of these parameters, Eqs. (12.84) and (12.82) are given by

$$q(t) = -\frac{1}{H^2(t)}\frac{\ddot{R}(t)}{R(t)} = \frac{8\pi G}{3H^2(t)}\left[\frac{1}{2}\rho(3\omega+1) - \rho_V\right]$$

$$= \left[\frac{1}{2}(3\omega+1)\Omega(t) - \Omega_V(t)\right]$$

$$\to \left[\frac{1}{2}(3\omega+1)\Omega(t) + \frac{1}{2}(3\omega_V+1)\Omega_V(t)\right], \tag{12.86}$$

$$kc^2 = R^2(t)H^2(t)\left[\Omega(t) + \Omega_V(t) - 1\right], \tag{12.87}$$

$$kc^2 = R_0^2 H_0^2\left[\Omega + \Omega_V - 1\right], \tag{12.88}$$

where $\Omega(t_0) = \Omega$.

Now the energy density ρ has several components as implied by the equation of the state

$$p_i = \omega_i \rho_i c^2 \tag{12.89}$$

corresponding to radiation, matter and vacuum. From Eq. (12.80)

$$\frac{\dot{\rho}_i}{\rho_i} = -3(1+\omega_i)\frac{\dot{R}}{R}. \tag{12.90}$$

Therefore, the solution of Eq. (12.90) gives

$$\frac{\dot{\rho}_i(t)}{\rho_i(0)} = \left(\frac{R(t)}{R_0}\right)^{-3(1+\omega_i)} = (1+z)^{3(1+\omega_i)}. \tag{12.91}$$

Now

$$\Omega(t) = \frac{\rho(t)}{\rho_c(t)} = \frac{\rho(t)}{\rho_0} \frac{\rho_0}{\rho_c(t_0)} \frac{\rho_c(t_0)}{\rho_c(t)} = \frac{\rho(t)}{\rho_0} \Omega \frac{H_0^2}{H^2(t)}$$

$$= (1+z)^{3(1+\omega)} \Omega \frac{H_0^2}{H^2(t)}.$$

Thus, we have

$$\Omega_i(t) = \frac{H_0^2}{H^2(t)} (1+z)^{3(1+\omega_i)} \Omega \qquad (12.92)$$

$$\Omega(t) = \frac{H_0^2}{H^2(t)} \sum_i (1+z)^{3(1+\omega_i)} \Omega_i. \qquad (12.93)$$

Hence from Eqs. (12.87), (12.88), we have

$$kc^2 = R^2(t) H^2(z) \left[\sum_i (1+z)^{3(1+\omega_i)} \Omega_i \frac{H_0^2}{H^2(z)} - 1 \right]$$

$$= \frac{R_0^2 H_0^2}{(1+z)^2} \left[\sum_i (1+z)^{3(1+\omega_i)} \Omega_i - \frac{H^2(z)}{H_0^2} \right]. \qquad (12.94)$$

Finally from Eqs. (12.94) and (12.88)

$$\frac{H^2(z)}{H_0^2} = \sum_i (1+z)^{3(1+\omega_i)} \Omega_i - \frac{kc^2(1+z)^2}{R_0^2 H_0^2} \qquad (12.95)$$

$$\frac{kc^2}{R_0^2 H_0^2} = (\Omega - 1) = \left(\sum_i \Omega_i - 1 \right) = (\Omega_r + \Omega_m + \Omega_V - 1) \quad (12.96)$$

and

$$q(t) \equiv q = \left[\frac{1}{2} \Omega_m(t) + \frac{3\omega_V + 1}{2} \Omega_V(t) \right] \qquad (12.97)$$

$$q_0 = \left[\frac{1}{2} \Omega_m + \frac{3\omega_V + 1}{2} \Omega_V \right] \qquad (12.98)$$

since Ω_r is negligible.

From Eq. (12.96), it follows that if

$$\Omega > 1, \ k > 0 : \text{Closed Universe}$$

$$\Omega < 1, \ k < 0 : \text{Open Universe}$$

$$\Omega = 1, \ k = 0 : \text{Spatially Flat Universe}.$$

Thus the sum of Ω_r, Ω_m and Ω_V, i.e., sum of densities determines the sign of the curvature. The current values of Ω's are (Ω_b: baryonic matter, Ω_{d_m}: dark matter)

$$\Omega_m = 0.27, \ \Omega_m = \Omega_b + \Omega_{dm}, \ \Omega_b \approx 0.045,$$

$$\Omega_{dm} \approx 0.22, \ \Omega_V \approx 0.73.$$

Thus

$$\Omega_r + \Omega_m + \Omega_V \approx 1 \tag{12.99}$$

then from Eq. (12.96), it follows that $k \approx 0$, i.e., the universe is flat.

From Eq. (12.98), it follows that if $w_V < -\frac{1}{3}$ then q_0 is negative, i.e., vacuum energy may lead to an accelerating universe. The vacuum energy is called the dark energy. The source of dark matter and dark energy are still unknown.

We conclude:

(1) For $w_V < -\frac{1}{3}$, the vacuum energy may lead to an accelerating Universe. Current data support it as $\Omega_m = 0.27$ and $\Omega_V \approx 0.73$. We see that Λ acts as "repulsive" (anti) gravity tending to speed up the expansion. It should be noted that it is the negative pressure of Λ rather than the energy density that does it.
(2) The presence of vacuum energy implies no link between geometry (curvature) and fate of the universe: (a) A closed universe ($k > 0$ which implies $\Omega_V > (1 - \Omega_m)$) can expand forever ($\Lambda$ is positive), (b) An open Universe ($k < 0$ which implies $\Omega_V \leq (1 - \Omega_m)$) must recollapse ($\Lambda$ is negative).

However, current data support $k = 0$, i.e., flat accelerating universe driven by negative pressure. From Eq. (12.73), it follows that for vacuum dominated universe

$$\frac{\ddot{R}}{R} = \frac{1}{3}\Lambda c^2 \tag{12.100}$$

and the integration gives

$$\frac{\dot{R}}{R} = \sqrt{\frac{1}{3}\Lambda c^2} = H$$

or

$$R(t) = R_0 e^{\sqrt{\frac{1}{3}\Lambda c^2}t} = R_0 e^{Ht}. \tag{12.101}$$

This gives the inflationary stage of the universe. Thus we see that curvature dominates at rather later time if Λ term does not dominate sooner.

If the first term dominates in Eq. (12.75), then

$$\frac{\dot{R}}{R} = \sqrt{\frac{8\pi G}{3}\rho} \sim R^{-\frac{3}{2}(1+\omega)} \text{ (cf. Eq. (12.81))}. \tag{12.102}$$

Integration gives

$$R^{\frac{3}{2}(1+\omega)} = t \tag{12.103}$$

$$R = t^{\frac{2}{3(1+\omega)}}.$$

Hence,

(i) matter dominated Universe, $\omega = 0$, $\rho \sim R^{-3}$, $R \sim t^{\frac{2}{3}}$.

(ii) radiation dominated Universe, $\omega = \frac{1}{3}$, $\rho \sim R^{-4}$, $R \sim t^{\frac{1}{2}}$.

(iii) for the curvature dominated Universe, it follows from Eq. (12.75) that

$$H^2 = \left(\frac{\dot{R}}{R}\right)^2 = \frac{|k|c^2}{R^2}, \quad R \sim t.$$

(iv) vacuum dominated Universe, $R \sim e^{\sqrt{\Lambda c^2/3}\,t} = e^{Ht}$.

Finally, from relations (12.54) and (12.53)

$$\frac{dR}{R_0} = \frac{H(z)}{1+z}dt, \tag{12.104}$$

$$\frac{dR}{R_0} = -\frac{1}{(1+z)^2}dz. \tag{12.105}$$

From the above two relations, one can write

$$dt = -\frac{dz}{(1+z)H(z)}. \tag{12.106}$$

For $t = t_0$, $R = R_0$, $z = 0$. Now $t \to 0$, as $R \to 0$, $z \to \infty$.

Hence from Eq. (12.106)

$$t_0 = \int_0^\infty \frac{dz}{(1+z)H(z)}. \tag{12.107}$$

On using Eqs. (12.95) and (12.96)

$$\frac{H^2(z)}{H_0^2} = \sum_i (1+z)^{3(1+\omega_i)}\,\Omega_i - (\Omega_r + \Omega_m + \Omega_V - 1)(1+z)^2. \tag{12.108}$$

Now $\omega_V = -1$, $\omega_m \to 0$, and neglecting Ω_r:

$$\frac{H^2(z)}{H_0^2} = (1+z)^2(\Omega_m z + 1) - \Omega_V z(2+z). \tag{12.109}$$

Hence, from Eqs. (12.107) and (12.109):

$$H_0 t_0 = \int_0^\infty \frac{dz}{(1+z)\left[(1+z)^2\left(\Omega_m z + 1\right) - \Omega_V z\left(2 + z\right)\right]^{1/2}}. \qquad (12.110)$$

For the special case of $\Omega_m + \Omega_V = 1$, $\Omega_m < 1$, one gets

$$H_0 t_0 = \int_0^\infty \frac{dz}{(1+z)\sqrt{\Omega_V}\left[1 + \frac{(1+z)^3(1-\Omega_V)}{\Omega_V}\right]^{1/2}}. \qquad (12.111)$$

By changing the variable

$$X^2 = \left[1 + \frac{(1+z)^3(1-\Omega_V)}{\Omega_V}\right].$$

The above integral can be simplified as

$$H_0 t_0 = \frac{2}{3}\frac{1}{\sqrt{\Omega_V}} \int_{1/\sqrt{\Omega_V}}^\infty \frac{dX}{X^2 - 1}$$

$$= \frac{1}{3\sqrt{\Omega_v}} \ln \frac{1 + \sqrt{\Omega_V}}{1 - \sqrt{\Omega_V}}. \qquad (12.112)$$

Using the observational values $\Omega_m = 0.27$, $\Omega_V = 0.73$, we see that $\Omega_m + \Omega_V = 1$, as used in Eq. (12.110). From Eq. (12.112):

$$H_0 t_0 \approx 0.99$$

$$t_0 \approx 13.7 \text{ Gyr},$$

i.e., age of Universe is 13.7 Gyr.

12.5   Problems

12.1 For the [2] sphere, embedded in [3], the metric is

$$ds^2 = R^2[d\theta^2 + \sin^2\theta d\varphi^2]. \qquad (12.113)$$

For the metric (12.113), i.e., for the [2] sphere embedded in [3], find:
(i) Ricci tensor and Ricci scalar
(ii) Riemannian tensor.

Solution

(i) The Christoffel symbols are given by (Problem 9.10)

$$\Gamma^\theta_{\phi\phi} = -R\sin\theta\cos\theta,$$

$$\Gamma^\phi_{\theta\phi} = \Gamma^\phi_{\phi\theta} = R\cot\theta,$$

$$\Gamma^\theta_{\theta\phi} = \Gamma^\theta_{\phi\theta} = 0,$$

$$\Gamma^\phi_{\phi\phi} = 0 = \Gamma^\theta_{\theta\theta}.$$

$$R_{\phi\phi} = \partial_\theta\Gamma^\theta_{\phi\phi} - \Gamma^\phi_{\phi\theta}\Gamma^\theta_{\phi\phi},$$

$$= R^2\sin^2\theta.$$

$$R_{\theta\theta} = -\partial_\theta\Gamma^\phi_{\theta\phi} - \Gamma^\phi_{\theta\phi}\Gamma^\phi_{\theta\phi},$$

$$= R^2.$$

$$R_{\theta\phi} = 0 = R_{\phi\theta}.$$

Ricci scalar

$$R^\mu_\mu = g^{\mu\nu}R_{\mu\nu},$$

$$R^0_0 + RR^\phi_\phi = g^{\theta\theta}R_{\theta\theta} + g^{\phi\phi}R_{\phi\phi} = 2.$$

(ii) Riemannian curvature tensor

$$R^\rho_{\mu\nu\lambda} = \partial_\nu\Gamma^\rho_{\mu\lambda} - \partial_\lambda\Gamma^\rho_{\mu\nu} + \Gamma^\sigma_{\mu\lambda}\Gamma^\rho_{\sigma\nu} - \Gamma^\sigma_{\mu\nu}\Gamma^\rho_{\sigma\lambda},$$

$$R^\theta_{\phi\theta\phi} = \partial_\theta\Gamma^\theta_{\phi\phi} - \partial_\theta\Gamma^\theta_{\phi\theta} + \Gamma^\sigma_{\phi\phi}\Gamma^\theta_{\sigma\theta} - \Gamma^\sigma_{\phi\theta}\Gamma^\theta_{\sigma\phi},$$

$$= \partial_\theta\Gamma^\theta_{\phi\phi} - \Gamma^\phi_{\phi\theta}\Gamma^\theta_{\phi\phi},$$

$$= R^2\sin^2\theta.$$

Curvature tensor

$$R_{\theta\phi\theta\phi} = g_{\theta\lambda}R^\lambda_{\phi\theta\phi},$$

$$= g_{\theta\theta}R^\theta_{\phi\theta\phi},$$

$$= R^2\sin^2\theta.$$

12.2 In Euclidean [3] space, the metric in spherical polar coordinates is

$$dS^2 = dr^2 + r^2(d\theta^2 + \sin^2\theta d\varphi^2).$$

Using the values of Christoffel symbols derived in Problem 9.9, show that

$$R_{\theta\theta} = R_{\varphi\varphi} = R_{\theta\varphi} = R_{\varphi\theta} = 0,$$

$$R^\theta_{\varphi\theta\varphi} = 0.$$

Thus the curvature tensors (Riemannian, Ricci) which distinguish the curvature of space from the curvature of curvilinear coordinate system of flat space signified by non-zero Christoffel symbols.

12.3 The metric for the [3] sphere embedded in [4] space:

$$dS^2 = R^2[d\chi^2 + \sin^2 x(d\theta^2 + \sin^2 \theta d\varphi^2)].$$

(i) Find the Christoffel symbols
(ii) Ricci tensor, Ricci scalar
(iii) Riemannian tensor, curvature tensor.

12.4 For the metric

$$d\tau^2 = a(r)dt^2 - [b(r)dr^2 + r^2(d\theta^2 + \sin^2 \theta d\varphi^2)].$$

Consider the motion of a particle of unit mass in the equatorial plane $\theta = \frac{\pi}{2}$. Show that the geodesic is given by

$$\frac{dt}{d\tau} = \frac{1}{a(r)},$$

$$r^2\frac{d\varphi}{d\tau} = J(\text{constant}),$$

$$\frac{dr^2}{d\tau^2} + \frac{a'(r)}{2b(r)a^2(r)} + \frac{b'(r)}{2b(r)}(\frac{dr}{d\tau})^2 - \frac{J^2}{r^3b(r)} = 0,$$

where $'$ means $\frac{\partial}{\partial r}$.

Hence show that

$$\frac{d}{d\tau}[b(r)(\frac{dr}{d\tau})^2 - \frac{1}{a(r)} + \frac{J^2}{r^2}] = 0,$$

which gives

$$b(r)(\frac{dr}{d\tau})^2 - \frac{1}{a(r)} + \frac{J^2}{r^2} = -E(\text{constant}),$$

or

$$\frac{b(r)}{a^2(r)}(\frac{dr}{dt})^2 - \frac{1}{a(r)} + \frac{J^2}{r^2} = -E.$$

Solution

$$g_{00} = a(r) = h_0,$$
$$g_{11} = -b(r) = h_1,$$
$$g_{22} = -r^2 = h_2,$$
$$g_{33} = -r^2 \sin^2 \theta = h_3.$$

From Eq. (12.15) of the text, the non-zero components of $\Gamma^\lambda_{\mu\nu}$ are

$$\Gamma^0_{10} = \frac{a'(r)}{2a(r)}, \ \Gamma^1_{00} = \frac{a'(r)}{2b(r)}, \ \Gamma^1_{11} = \frac{b'(r)}{2b(r)},$$

$$\Gamma^1_{22} = -\frac{r}{b(r)}, \ \Gamma^1_{33} = -\frac{r\sin^2\theta}{b(r)}, \tag{12.114}$$

$$\Gamma^2_{12} = \frac{1}{r}, \ \Gamma^3_{13} = \frac{1}{r}, \ \Gamma^3_{23} = \cot\theta.$$

Geodesic equations:

$$\frac{d^2x^\mu}{d\tau^2} + \Gamma^\mu_{\nu\lambda}\frac{dx^\nu}{d\tau}\frac{dx^\lambda}{d\tau} = 0.$$

Using Eq. (12.114) ($\theta = \frac{\pi}{2}$), the geodesic equations are:

$$x^0 = t,$$

$$\frac{d^2t}{d\tau^2} + 2\Gamma^0_{10}\frac{dx^1}{d\tau}\frac{dx^0}{d\tau} = 0,$$

or

$$\frac{d^2t}{d\tau^2} + \frac{a'(r)}{a(r)}\frac{dr}{d\tau}\frac{dt}{d\tau} = \frac{d}{d\tau}[\ln(\frac{dt}{d\tau}) + \ln a(r)] = 0.$$

Thus

$$\frac{dt}{d\tau} = \frac{\text{constant}}{a(r)}, \tag{12.115}$$

$$\frac{d^2x^3}{d\tau^2} + \Gamma^3_{\nu\lambda}\frac{dx^\nu}{d\tau}\frac{dx^\lambda}{d\tau} = 0,$$

or

$$\frac{d^2\phi}{d\tau^2} + 2\Gamma^3_{13}\frac{dr}{dt}\frac{d\phi}{d\tau} = 0,$$

or

$$\frac{d^2\phi}{d\tau^2} + \frac{2}{r}\frac{dr}{d\tau}\frac{d\phi}{d\tau} = \frac{d}{d\tau}[\ln\frac{d\phi}{d\tau} + \ln r^2] = 0.$$

Hence we have

$$r^2\frac{d\phi}{d\tau} = J(\text{constant}) \tag{12.116}$$

$$\frac{d^2x^1}{d\tau^2} + \Gamma^1_{\nu\lambda}\frac{dx^\nu}{d\tau}\frac{dx^\lambda}{d\tau} = 0,$$

$$\frac{d^2r}{d\tau^2} + \Gamma^1_{00}(\frac{dt}{d\tau})^2 + \Gamma^1_{11}(\frac{dr}{d\tau})^2 + \Gamma^1_{33}(\frac{d\phi}{d\tau})^2 = 0.$$

Thus

$$\frac{d^2r}{d\tau^2} + \frac{\acute{a}(r)}{2b(r)a^2(r)} + \frac{\acute{b}(r)}{2b(r)}(\frac{dr}{d\tau})^2 - \frac{1}{b(r)}\frac{J^2}{r^3} = 0.$$

The above equation can be put in the form

$$\frac{d}{d\tau}[b(r)(\frac{dr}{d\tau})^2 - \frac{1}{a(r)} + \frac{J^2}{r^2}] = 0,$$

so that

$$[b(r)(\frac{dr}{d\tau})^2 - \frac{1}{a(r)} + \frac{J^2}{r^2}] = -E(\text{constant}),$$

or

$$\frac{b(r)}{a^2(r)}(\frac{dr}{dt})^2 - \frac{1}{a(r)} + \frac{J^2}{r^2} = -E, \qquad (12.117)$$

on using

$$\frac{dr}{d\tau} = (\frac{dr}{dt})\frac{dt}{d\tau} = \frac{1}{a(r)}\frac{dr}{dt}.$$

A good approximation to the metric outside the earth is provided by

$$a(r) = (1 + 2\Phi),$$

$$b(r) = \frac{1}{1 + 2\Phi} \simeq 1 - 2\Phi, \quad \Phi(r) = -\frac{GM}{r}.$$

For this metric, Eq. (12.117) gives for a slowly moving particle in a weak gravitation field ($\Phi \ll 1$)

$$r^2\frac{d\Phi}{dt} = J,$$

$$\frac{1}{2}\dot{r}^2 + \frac{J^2}{2r} + \Phi \approx \frac{1 - E}{2}.$$

Consider a clock on the surface of the earth a distance r, from the centre of the earth and clock at a tall tower at a distance r_r from the centre of the earth. Calculate the time elapsed on each clock. From Eq. (12.115)

$$\frac{dt}{d\tau} = \frac{1}{1 + 2\Phi}.$$

Thus

$$t \approx \tau(1 - 2\Phi)$$

$$= \tau\left(1 + \frac{2GM}{r}\right)$$

Hence we have

$$t_1 - t_2 = 2\tau GM\frac{r_2 - r_1}{r_1 r_2}$$

$$t_1 > t_2$$

the clock at the tower is slow.

12.5 Energy conservation

$$\nabla_\mu T_0^\mu = 0.$$

Derive the equation

$$\dot{\rho} = -\frac{3H}{c^2}(\rho c + p).$$

12.6 Derive the equations describing the matter dominated Universe for the three cases $k = 1$, $k = -1$ and $k = 0$.

Solution

For the matter dominated universe

$$\rho R^3 = \text{constant} = \frac{3}{4\pi}M,$$

where M is a constant with dimensions of mass (in other words the mass of co-moving region remains constant).

Introduce cosmic time η:

$$dt = R d\eta$$

$$\frac{dR}{dt} = \frac{1}{R}\frac{dR}{d\eta},$$

Eq. (12.75), we have ($\Lambda = 0$, $c = 1$)

$$\dot{R}^2 = \frac{8\pi G}{3}R^2\rho - k \qquad (12.118)$$

or from Eq. (12.118)

$$\left(\frac{dR}{d\eta}\right)^2 = \frac{8\pi G}{3}\rho R^4 - kR^2$$

$$\frac{dR}{d\eta} = \sqrt{2GMR - R^2}$$

(i) For $k = 1$, by change of variable

$$R = 2GM \sin^2\theta$$

we obtain $\theta = \frac{\eta}{2}$, which implies

$$\sin\theta = \sin\frac{\eta}{2}.$$

Hence, for $k = 1$ (positively curved Universe)

$$R = GM(1 - \cos\eta),$$

and using

$$dt = Rd\eta$$

$$t = GM\left(\eta - \sin\eta\right).$$

(ii) Similarly for $k = -1$ (negatively curved Universe)

$$R = GM\left(\cosh\eta - 1\right)$$

$$t = GM\left(\sinh\eta - \eta\right).$$

(iii) For $k = 0$, flat Universe,

$$\frac{dR}{d\eta} = \sqrt{2GMR}.$$

Thus

$$R^{1/2} = \sqrt{\frac{GM}{2}}\eta \Rightarrow R = \frac{GM}{2}\eta^2$$

$$\frac{dt}{d\eta} = R = \frac{GM}{2}\eta^2$$

$$t = \frac{GM}{6}\eta^3 = \frac{GM}{6}\left(\frac{2R}{GM}\right)^{3/2}.$$

Thus

$$R = \left(\frac{9GM}{2}\right)^{1/3} t^{2/3}.$$

To conclude: For $k = 1$, the Universe will recollapse in a finite time, whereas for $k = 0, -1$, the Univese will expand indefinitely. These simple conclusion can be altered when $\Lambda \neq 0$. Thus in the presence of Λ, there is no link between geometry (curvature) and fate of the Universe.

APPENDIX

A.1 Covariant Derivative; Parallel Displacement of a Vector

The covariant derivative in curved space (Riemann's geometry) has been discussed in sec. 9.3. In this appendix, the covariant derivative in the context of parallel displacement of a vector is discussed. As already noted for a scalar field $f(x)$ $[dx^\lambda = \varepsilon n^\lambda]$

$$\lim_{\epsilon \to 0} \frac{f(x^\lambda + \varepsilon n^\lambda) - f(x^\lambda)}{\epsilon} = \partial_\lambda f = \frac{\partial f}{\partial x^\lambda}, \qquad (A.1)$$

is a covariant vector. For a vector field V^μ

$$V'^\mu(x') = V^\nu(x) \frac{\partial x'^\mu}{\partial x^\nu}. \qquad (A.2)$$

The important point to note is that definition like (A.1) viz

$$dx^\lambda \partial_\lambda V^\mu(x) = V^\mu(x^\lambda + dx^\lambda) - V^\mu(x^\lambda), \qquad (A.3)$$

does not make sense since $V^\mu(x^\lambda + dx^\lambda)$ transforms differently from $V^\mu(x^\lambda)$. It is necessary to define a parallel displacement of a vector V^μ from x^λ to $x'^\lambda = x^\lambda + \varepsilon n^\lambda$, $dx^\lambda = \varepsilon n^\lambda$ and denote it by $V'^\mu(x^\lambda + \varepsilon n^\lambda)$ [see Fig. A.1]. The definition of covariant derivative then naturally follows

$$n^\lambda \nabla_\lambda V^\mu(x) = \lim_{\varepsilon \to 0} \frac{V^\mu(x^\lambda + \varepsilon n^\lambda) - V'^\mu(x^\lambda + \varepsilon n^\lambda)}{\varepsilon}. \qquad (A.4)$$

Now using Taylor's expansion

$$V^\mu(x^\lambda + \varepsilon n^\lambda) = V^\mu(x^\lambda) + \varepsilon n^\lambda \frac{\partial V^\mu}{\partial x^\lambda} \qquad (A.5)$$

and from transformation law (A.2)

$$V'^{\mu}(x^{\lambda} + \varepsilon n^{\lambda}) = V^{\nu}(x^{\lambda})[\frac{\partial x^{\mu}}{\partial x^{\nu}} + \varepsilon \frac{\partial n^{\mu}}{\partial x^{\nu}}]$$

$$= V^{\mu}(x^{\lambda}) - \varepsilon n^{\lambda}\Gamma^{\mu}_{\lambda\nu}V^{\nu}, \qquad (A.6)$$

where we have used

$$\frac{\partial n^{\mu}}{\partial x^{\nu}} = -n^{\lambda}\Gamma^{\mu}_{\lambda\nu}, \qquad \Gamma^{\mu}_{\lambda\nu} = \Gamma^{\mu}_{\nu\lambda}$$

so that derivative of normal satisfies linear relation.

Substituting Eq. (A.5) and Eq. (A.6) into Eq. (A.4), we obtain

$$n^{\lambda}\nabla_{\lambda}V^{\mu}(x) = n^{\lambda}\partial_{\lambda}V^{\mu} + n^{\lambda}\Gamma^{\mu}_{\lambda\nu}V^{\nu}. \qquad (A.7)$$

Since n^{λ} is arbitrary,

$$\nabla_{\lambda}V^{\mu} = \partial_{\lambda}V^{\mu} + \Gamma^{\mu}_{\lambda\nu}V^{\nu} = V^{\mu}_{;\lambda}. \qquad (A.8)$$

Roughly speaking, the first term comes from change in vector field from x^{λ} to $x^{\lambda} + dx^{\lambda}$ and second from change in basic vectors x^{λ}.

Remarks

1- In case $\Gamma^{\mu}_{\nu\lambda} = 0$

$$V'^{\mu}(x^{\lambda} + \varepsilon n^{\lambda}) = V^{\mu}(x^{\lambda}) \Rightarrow \frac{\partial V^{\mu}}{\partial x^{\lambda}} = 0. \qquad (A.9)$$

This is the case for flat space, with inertial coordinates where the propagation of vector field along a curve with constant components is possible.

From Eq. (A.4) the right hand side being difference of two vectors at the same point, it is clear that $x^{\lambda}\nabla_{\lambda}V^{\mu}$ is a vector and from quotient law $\nabla_{\lambda}V^{\mu}$ is a second rank mixed tensor.

2- In order to get more insight, it is helpful to introduce the notation of parallel transport of a vector along a path keeping its component constants. It is straightforward in flat space where a vector field V^{μ} with constant components is represented by a field of parallel vectors along a curve. Thus parallel transport of vector in flat space from one point to another point along a curve $x^{\mu}(s)$ is expressed

$$\frac{dV^{\mu}}{ds} = \frac{\partial V^{\mu}}{\partial x^{\lambda}}\frac{dx^{\lambda}}{ds} = 0. \qquad (A.10)$$

However, in curved space, the parallel transport of a vector from point $P(x^{\lambda})$ to point $P(x^{\lambda} + dx^{\lambda})$ depends on the components themselves but the dependence must be linear viz

$$dV^{\mu} = -\Gamma^{\mu}_{\lambda\nu}V^{\nu}dx^{\lambda}.$$

Hence for a parallel transport of a vector from a point $P(s)$ to $P(s+ds)$ along a curve $x^\lambda(s)$, we get

$$\frac{dV^\mu}{ds} + \Gamma^\mu_{\lambda\nu} V^\nu \frac{dx^\lambda}{ds} = 0. \tag{A.11}$$

This is the equation of parallel transport of a vector V^μ in curved space. The importance of parallel displacement of a vector in the context of Riemann's geometry was emphasized by Herman Weyl.

A.2 Hot Big Bang: Thermal History of the Universe

A.2.1 *Thermal Equilibrium*

Consider an arbitrary volume V in thermal equilibrium with a heat bath at temperature T. The particle density n_i (i, particle index) at temperature T is given by

$$n_i = \frac{N_i}{V} = \frac{g_i}{2\pi^2} \left(\frac{k_B T}{\hbar c}\right)^3 \int_0^\infty \left[\exp\left(\frac{E}{k_B T}\right) \pm 1\right]^{-1} z^2 dz. \tag{A.12}$$

The energy density is given by

$$\rho_i c^2 = \frac{g_i}{2\pi^2} \left(\frac{k_B T}{\hbar c}\right)^3 (k_B T) \int_0^\infty \left[\exp\left(\frac{E}{k_B T}\right) \pm 1\right]^{-1} \left(\frac{E}{k_B T}\right) z^2 dz, \tag{A.13}$$

where

$$z = \frac{qc}{k_B T}, \qquad E = \left[(q_i c)^2 + (m_i c^2)^2\right]^{1/2} \tag{A.14}$$

and g_i are the number of spin states, q_i is the momentum of the particle and m_i is its mass. The $+$ sign is for the fermions (F) and $-$ sign is for the bosons (B). In particular for $i = $ photon, $m = 0$, $g = 2$. In writing Eqs. (A.13) and (A.14), we have put the chemical potential $\mu_i = 0$ for photons. Since particles and antiparticles are in equilibrium with photons $\mu_i = -\mu_{\bar{i}}$, there is no asymmetry between the number of particles and antiparticles, $\mu_i = \mu_{\bar{i}} = 0$. Even if that is not true, the difference between the number of particles and antiparticles is small compared with the number of photons,

$$\left|\frac{\mu_i}{k_B T}\right| = \left|\frac{\mu_{\bar{i}}}{k_B T}\right| \ll 1 \tag{A.15}$$

and the chemical potential can be neglected. For the photon gas, we get from Eqs. (A.12), (A.13) and (A.14)

$$n_\gamma = 2\frac{\zeta(3)}{\pi^2} \left(\frac{k_B T}{\hbar c}\right)^3 = 2\frac{1.2}{\pi^2} \left(\frac{1}{\hbar c}\right)^3 (k_B T)^3 \tag{A.16}$$

$$\rho_\gamma \, c^2 = 6 \frac{\zeta(4)}{\pi^2} \left(\frac{1}{\hbar c} \right)^3 (k_B T)^4$$

$$= \frac{\pi^2}{15} \left(\frac{1}{\hbar c} \right)^3 (k_B T)^4 \approx 2.7 \, n_\gamma \, (k_B T), \qquad \text{(A.17)}$$

where we have used

$$\int_0^\infty \frac{z^2}{e^z - 1} dz = \Gamma(3)\zeta(3) = 2 \times 1.2$$

$$\int_0^\infty \frac{z^3}{e^z - 1} dz = \Gamma(4)\zeta(4) = 6 \times \zeta(4)$$

$$= \frac{(2\pi)^4}{8} \frac{1}{30}$$

$\zeta(3)$ and $\zeta(4)$ are Riemann zeta functions. For a gas of extreme relativistic particles (ER), $k_B T \gg m_i c^2$, $qc \gg m_i c^2$, we thus get

$$n_B = \left(\frac{g_B}{2} \right) n_\gamma, \qquad \rho_B = \left(\frac{g_B}{2} \right) \rho_\gamma \qquad \text{(A.18)}$$

$$n_F = \frac{3}{4} \left(\frac{g_F}{2} \right) n_\gamma, \qquad \rho_F = \frac{7}{8} \left(\frac{g_F}{2} \right) \rho_\gamma. \qquad \text{(A.19)}$$

The entropy S for the photon gas is given by

$$S = \frac{R^3}{T} \frac{4}{3} \, \rho_\gamma \, (T) \, c^2. \qquad \text{(A.20)}$$

For any relativistic gas

$$S = \frac{R^3}{T} \frac{4}{3} \, \rho \, (T) \, c^2. \qquad \text{(A.21)}$$

Thus for a gas consisting of extreme relativistic particles (bosons and fermions):

$$n(T) = \frac{1}{2} \, g'(T) \, n_\gamma(T)$$

$$= \frac{1.2}{\pi^2} \, g'(T) \left(\frac{k_B T}{\hbar c} \right)^3 \qquad \text{(A.22)}$$

$$\rho \, (T) \, c^2 = \frac{1}{2} \, g_*(T) \, \rho_\gamma(T) c^2$$

$$= \frac{\pi^2}{30} \, g_*(T) \left(\frac{k_B T}{\hbar c} \right)^3 (k_B T) \qquad \text{(A.23)}$$

$$S = \frac{R^3}{T} \frac{2}{3} \, g_*(T) \, \rho_\gamma(T) c^2, \qquad \text{(A.24)}$$

where

$$g'(T) = \sum_B g_B + \frac{3}{4} \sum_F g_F \qquad (A.25)$$

$$g_*(T) = \sum_B g_B + \frac{7}{8} \sum_F g_F \qquad (A.26)$$

are called the "effective" degrees of freedom. We note that entropy per unit volume is given by

$$\frac{s}{k_B} \equiv \frac{1}{k_B} \frac{S}{R^3} = \frac{2\pi^2}{45} g_*(T) \left(\frac{k_B T}{\hbar c} \right)^3. \qquad (A.27)$$

For non-relativistic gas $k_B T \ll m_i c^2$, we use the Boltzmann distribution

$$n_i = \frac{g_i}{2\pi^2} \left(\frac{k_B T}{\hbar c} \right)^3 \int_0^\infty \exp \left(-\frac{E}{k_B T} \right) z^2 \, dz \qquad (A.28)$$

$$E \approx m_i c^2 \left[1 + \frac{1}{2} \frac{q^2 c^2}{(m_i c^2)^2} \right]. \qquad (A.29)$$

From Eq. (A.28), we get

$$n_i = \left[\frac{g_i}{(2\pi)^{3/2}} \right] \left(\frac{k_B T}{\hbar c} \right)^3 \left[\left(\frac{m_i c^2}{k_B T} \right)^{3/2} e^{-m_i c^2 / k_B T} \right] \qquad (A.30)$$

$$\rho_i = n_i \, m_i. \qquad (A.31)$$

A.2.2 The Radiation Era

For extreme relativistic gas, $\rho = \frac{1}{3} \rho c^2$, we get from Eq. (12.78b)

$$R^3 \frac{d\rho}{dR} + 4\rho R^2 = 0. \qquad (A.32)$$

Thus, we have

$$\rho = A^2 R^{-4} \qquad (A : \text{constant}). \qquad (A.33)$$

For non-relativistic gas $p \approx 0$, we have

$$R^3 \frac{d\rho}{dR} + 3\rho R^2 = 0$$

$$\rho_{N.R} = (\text{constant}) R^{-3} \qquad (A.34)$$

Hence

$$\frac{\rho_{N \cdot R}}{\rho_{E \cdot R}} \propto R \to 0 \qquad \text{as} \quad R \to 0.$$

Therefore, we have the important result. Early universe is dominated by extreme relativistic particles, i.e., the universe is radiation dominated in early stages. Since $\rho \to \frac{1}{R^4}$ for the early universe, we can neglect the second and third terms on the right-hand side of Eq. (12.75) as compared with the first term. Thus we get

$$H = \frac{\dot{R}}{R} = \sqrt{\frac{8\pi}{3}} \, (G_N \, \rho)^{1/2} . \tag{A.35}$$

Now using Eq. (A.23),

$$H = \sqrt{\frac{4\pi^3}{45}} \, [g_* \, (T)]^{1/2} \, \frac{(k_B T)^2}{(M_P c^2)\hbar} \approx 0.21 \, g_*^{1/2} \left(\frac{k_B T}{\text{MeV}}\right)^2 \text{sec}^{-1} . \tag{A.36}$$

where M_P is the Planck mass $= \sqrt{\frac{\hbar c}{G_N}}$. Further, using Eq. (A.33), we get from Eq. (A.35)

$$R\dot{R} = A\sqrt{\frac{8\pi G_N}{3}} . \tag{A.37}$$

Hence

$$t = \sqrt{\frac{3}{32\pi G_N}} \frac{R^2}{A} = \sqrt{\frac{3}{32\pi}} \frac{1}{\sqrt{G_N \, \rho}}$$

$$= \sqrt{\frac{45}{16\pi^3}} \, g_*^{-1/2} \, \frac{\hbar(M_P c^2)}{(k_B T)^2}$$

$$= 2.42 \, g_*^{-1/2} \left(\frac{\text{MeV}}{k_B T}\right)^2 \text{sec}. \tag{A.38}$$

Thus from Eq. (A.36) and (A.38)

$$Ht \approx 0.5. \tag{A.39}$$

We consider two examples:
  (i) For $g_* = g_\gamma = 2$,

$$t \approx 1.7 \left(\frac{\text{MeV}}{k_B T}\right)^2 \text{sec}. \tag{A.40}$$

Thus for $k_B T = m_e c^2 \approx 0.51$ MeV, $t = 6.5$ sec and $H \approx 0.08$ sec^{-1}.
  (ii) For $m_\mu > k_B T > m_e$,

$$g_* = g_\gamma + \frac{7}{8}(g_e + 3g_\nu)$$

$$= 2 + \frac{7}{8}(4 + 6) = \frac{43}{4} \tag{A.41}$$

and for $m_\pi > k_B T > m_\mu$,

$$g_* = 2 + \frac{7}{8} \left(g_e + g_\mu + 3g_\nu \right)$$

$$= \frac{57}{4}. \tag{A.42}$$

Here we have taken the number of neutrinos $N_\nu = 3$. Now we get from Eqs. (A.35) and (A.37) at $k_B T = 1$ MeV

$$H \approx 0.69 \left(\frac{k_B T}{\text{MeV}} \right)^2 \ \sec^{-1} \approx 0.69 \ \sec^{-1} \tag{A.43}$$

and

$$t \approx 0.74 \left(\frac{\text{MeV}}{k_B T} \right)^2 \ \sec \approx 0.74 \ \sec. \tag{A.44}$$

Now Eq. (A.44) gives the time evolution of the universe in radiation era. From Eq. (12.75), we have the important result that $H \propto \sqrt{\rho}$, i.e., the higher the energy density in the early universe, the faster will be the expansion rate.

As we have seen, the radiation density falls off as R^{-4} and the energy density in non-relativistic matter falls of as R^{-3}. The universe eventually becomes matter dominated. At $t = t_{eq}$, matter density becomes equal to radiation density, i.e.,

$$\rho_m \left(t_{eq} \right) = \rho_r \left(t_{eq} \right), \tag{A.45}$$

where on using Eq. (12.87)

$$\rho_m \left(t_{eq} \right) = \Omega_{mo} \, \rho_{co} \left(\frac{R_o}{R_{eq}} \right)^3 \tag{A.46}$$

and from Eqs. (A.17), (A.19) and (A.33) we get

$$\rho_m \left(t_{eq} \right) = \rho_r \left(t_{eq} \right) = \frac{\pi^2}{30} \left[g_\gamma \, (k_B T_{\gamma o})^4 + \frac{7}{8} 3 \, g_\nu \, (k_B T_{\nu o})^4 \right] \frac{1}{(\hbar c)^3} \frac{1}{c^2}$$

$$\times \left(\frac{R_o}{R_{eq}} \right)^4$$

$$= \frac{\pi^2}{30} \left[2 + \frac{21}{4} \left(\frac{4}{11} \right)^{4/3} \right] \frac{(k_B T_0)^4}{(\hbar c)^3} \left(\frac{R_o}{R_{eq}} \right)^4$$

$$= \frac{1.106}{c^2} \frac{(k_B T_0)^4}{(\hbar c)^3} \left(\frac{R_o}{R_{eq}} \right)^4 \tag{A.47}$$

where we have used $\left(\frac{T_{\nu_o}}{T_{\gamma_o}}\right)^3 = \frac{4}{11}$, [see Eq. (18.135) in chapter 18 of [14] and $T_{\gamma_o} = T_0$. Hence from Eqs. (A.46) and (A.47), we obtain for $T_0 = 2.725 K$ [cf. Eq. (12.53)]

$$1 + z_{eq} = \frac{R_o}{R_{eq}} = \frac{1}{1.106}\frac{(\hbar c)^3}{(k_B T_0)^4}\Omega_{mo}(\rho_{co}c^2)$$

$$= (2.28 \times 10^6 \text{ MeV}^{-1}\text{-cm}^3)\Omega_{mo}(\rho_{co}c^2)$$

where

$$\rho_c c^2 = \frac{3}{8\pi}\frac{H^2}{G_N}c^2$$

$$= \frac{3}{8\pi}H^2(M_P c^2)\frac{1}{(\hbar c^3)}$$

$$= h^2(10.054^{-13} \text{ MeV-cm}^{-3}).$$

Thus

$$1 + z_{eq} = 2.29 \times 10^4\Omega_{mo}h_o^2.$$

Now

$$\Omega_{mo} = 0.27, \quad h_o = 0.710$$

$$\Omega_{mo}h_o^2 = 0.136 \rightarrow 0.113$$

$$1 + z_{eq} \simeq 3114 \rightarrow 3201 \text{ (observed value)}. \qquad (A.48)$$

From Eqs. (A.47) and (A.48), we get

$$k_B T_{eq} = (k_B T_0)\frac{R_0}{R_{eq}} = (k_B T_0)(1 + z_{eq})$$

$$\simeq 0.75 \text{ eV}. \qquad (A.49)$$

From Eqs. (A.37) and (A.47), we obtain

$$t_{eq} = \sqrt{\frac{3}{32\pi}}\frac{1}{\sqrt{G_N\rho_{eq}}}$$

$$= \sqrt{\frac{3}{(1.106)32\pi}}\frac{\hbar(M_P c^2)}{(k_B T_0)^2}\left(\frac{R_o}{R_{eq}}\right)^{-2}$$

$$\simeq 0.164\frac{\hbar(M_P c^2)}{(k_B T_0)^2}\frac{1}{(1 + z_{eq})^2}$$

$$\simeq 2.33 \times 10^{12} \text{ sec}.$$

In the dense early universe the radiation would have been held in thermal equilibrium with matter and would have scattered repeatedly off free electrons. But when the expansion had cooled the matter below 3000 K ($k_B T \simeq 0.26$ eV), so that from Eq. (A.38) with $g_* = 2 + \frac{7}{8}(6) = \frac{29}{4}$, $t \simeq 1.3 \times 10^{13}$ sec $\simeq 4 \times 10^5$ yrs, the primordial plasma would have recombined with atoms, the universe thereafter becoming transparent to light. The experimentally detected microwave photons are therefore direct messengers from an era when the universe had an age of about 4×10^5 yrs. But photons are still around and they fill the universe and have nowhere else to go. The thermal radiation last scattered at this epoch is now detected as the cosmic background radiation. This epic, which corresponds $1 + z_{eq} \simeq 1100$, defines a "surface" known as "surface of last scattering".

Era	Age (in seconds)	Temperature K	Remarks
Big Bang	0		Vacuum to matter transition
End of Inflation	10^{-36}	10^{27}	
Electro-weak	10^{-10}	10^{18}	$W^\pm, Z^0$: Electroweak transition
Quark	10^{-5}	3×10^{12}	Hadronization
Lepton	8×10^{-5}	1.2×10^{12}	$\mu^\pm$ annihilation
	8×10^{-3}	1.2×10^{11}	ν_μ's decouple
	0.7	10^{10}	ν_e's decouple
	6	6×10^9	$e^+ e^-$ annihilation
Particle	100	$[1.3 - 0.8] \times 10^9$	Nucleosynthesis
Photon	2×10^{12}	2×10^{14}	Radiation era ends, *CMB* decouples
	10^{13}	3×10^3	from plasma
Atoms	10^{14}	10^3	Plasma to atom;
Now	4×10^{17}	2.75	Present

We summarize the cosmic history of the universe in the Table above. In particular

$$\rho_{\text{radiation}} = \rho_{\text{matter}}$$

$$z_{eq} = 3200; \ t_u \simeq 7.4 \times 10^4 \text{ yrs}$$

CMB decouples from Plasma

$$z_{\text{dec}} = 1089, \ t_u = 380,000 \text{ yrs}$$

$$T_{CMB} = 2970 \text{ K}.$$

First stars are thought to form at

$$z_r = 20, \ t_u = 2 \times 10^8 \text{ yrs}.$$

Now

$$z = 0, \ t_u = 13.7 \times 10^9 \text{ yrs}$$

$$T_{CMB} = 2.725 \text{ K}.$$

We end this section by writing some useful numbers. From Eqs. (A.16) and (A.17), using the present temperature $T_0 = 2.735$ K, we get

$$n_{\gamma_0} \approx \frac{2.4}{\pi^2} \left(\frac{k_B \, T_0}{\hbar c} \right)^3 \approx 415 \text{ cm}^{-3} \qquad (\text{A.50})$$

$$\rho_{\gamma_0} \approx 2.6 \times 10^{-10} \text{ GeV}/c^2 \text{ cm}^{-3}. \qquad (\text{A.51})$$

Thus n_γ at temperature T is given by

$$n_\gamma \approx 415 \left(\frac{T}{2.725} \right)^3 \text{ cm}^{-3}. \qquad (\text{A.52})$$

A.3   Fundamental Units

Speed of light:

$$c = 2.998 \times 10^8 \text{ m sec}^{-1}$$

$$= 2.998 \times 10^{10} \text{ cm sec}^{-1}$$

$$\approx 3 \times 10^{10} \text{ cm sec}^{-1}$$

Planck's constant:

$$\hbar = \frac{h}{2\pi} \approx 6.582 \times 10^{-22} \text{ MeV-sec}$$

$$= 6.582 \times 10^{-25} \text{ GeV-sec}$$

$$\hbar c \approx 1.973 \times 10^{-11} \text{ MeVcm}$$

$$= 1.973 \times 10^{-16} \text{ GeVm} \qquad (\text{A.53})$$

Electric Charge:

$$e = 1.602 \times 10^{-19} \text{ C}$$

$$\frac{1}{4\pi\epsilon_0} = 8.987 \times 10^9 \text{ Jm} - \text{C}^{-2}$$

$$1\text{eV} = 1.602 \times 10^{-19} \text{ J}$$

Fine structure constant:

$$\alpha = \frac{e^2}{\hbar c}\frac{1}{4\pi\epsilon_0} \approx \frac{1}{137}$$

Mass of electron:

$$m_e \approx 9.109 \times 10^{-31} \text{ kg}$$

$$m_e c^2 = 9.109 \times 10^{-31}(9 \times 10^{16} \text{ J})$$

$$= 8.198 \times 10^{-14} \text{ J}$$

$$\approx 0.511 \text{ MeV}$$

$$m_e = 0.511 \text{ MeV}/c^2$$

Radius of electron:

$$r_e = \alpha\frac{e\hbar}{m_e c^2} \approx 2.82 \times 10^{-15} \text{ m}$$

$$= 2.82 \times 10^{-13} \text{ cm}$$

Fermi constant:

$$\frac{G_F}{(\hbar c)^3} \approx 1.66 \times 10^{-5} \text{ GeV}^{-2}$$

Newton's gravitational constant:

$$G_N = 6.673 \times 10^{-11} \text{ m}^3\text{kg}^{-1}\text{sec}^{-2}$$

$$= 6.673 \times 10^{-11} \text{ Jkg}^{-2}\text{m}$$

$$= 1.323 \times 10^{-54} \text{ GeV-m}\left(\frac{\text{GeV}}{c^2}\right)^{-2}$$

$$\frac{G_N}{\hbar c} = 6.70 \times 10^{-39}\left(\frac{\text{GeV}}{c^2}\right)^{-2}$$

$$\sqrt{\frac{\hbar c}{G_N}} = 1.221 \times 10^{19} \text{ GeV}/c^2 = M_P \text{ Planck mass}$$

Planck radius:

$$\frac{\hbar c}{M_P c^2} = 1.616 \times 10^{-35} \text{ m}$$

Boltzmann constant:
$$k_B = 1.380 \times 10^{-23} \text{ J K}^{-1}$$
$$= 8.617 \times 10^{-11} \text{ MeV K}^{-1}$$
$$= 8.617 \times 10^{-5} \text{ eV K}^{-1}$$

Present CMB temperature:
$$T_o = 2.725 \text{ K}$$
$$k_B T_o = 2.348 \times 10^{-10} \text{ MeV}$$
$$= 2.348 \times 10^{-4} \text{ eV}$$

Hubble parameter:
$$H = h100 \text{ km sec}^{-1}\text{Mpc}^{-1}$$
$$\text{Parsec: } pc = 3.0856 \times 10^{13} \text{ km}$$
$$1 \text{ Mpc} = 3.0856 \times 10^{19} \text{ km}$$
$$\text{Mpc}^{-1} = 3.241 \times 10^{-20} \text{ km}^{-1}$$
$$H = h32.409 \times 10^{-19} \text{ sec}^{-1}$$
$$H^{-1} = h^{-1}3.0856 \times 10^{17} \text{ sec}$$

Present value of h:
$$h_o = 0.710$$

Present value of Hubble parameter:
$$H_o = 22.755 \times 10^{-19} \text{ sec}^{-1}$$
$$H_o^{-1} = 4.3459 \times 10^{19} \text{ sec}$$
$$1 \text{ year} = 3.1558 \times 10^7 \text{ sec}$$
$$H_o^{-1} = 1.377 \times 10^{12} \text{ yr}$$
$$= 1.377 \text{ Gyr}$$
$$1 \text{ light year} = 2.998 \times 10^8 \text{ sec}^{-1}(3.1558 \times 10^7 \text{ sec m})$$
$$= 9.461 \times 10^{15} \text{ m}$$
$$= 9.461 \times 10^{12} \text{ km}$$

Natural Units:
$$\hbar = c = 1$$
$$L = \frac{1}{\text{GeV}} = \frac{\hbar c}{\text{GeV}} = 1.973 \times 10^{-16} \text{ m}$$
$$T = \frac{1}{\text{GeV}} = \frac{\hbar}{\text{GeV}} = 6.582 \times 10^{-25} \text{ sec}$$

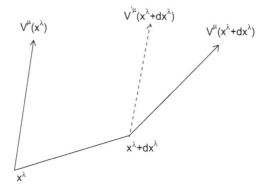

Fig. A.1   Parallel transport of a vector.

Bibliography

1. *Theory of Relativity*, W. Pauli, Dover Publications, 1958.
2. *Space-Time Matter*, H. Weyl, Dover Publications (June 1, 1952).
3. *Classical Theory of Fields*, L. Landau and E. M. Lifshitz, Butterworth-Heinemann; 4th edition (December 31, 1975).
4. *Gravitation and Cosmology: Principles and Applications of General Relativity*, S. Weinberg, John Wiley & Sons, Inc.; 1st edition (July 1972).
5. *Special Relativity (The M.I.T. Introductory Physics Series)*, A. P. French, W. W. Norton & Company (August 17, 1968).
6. *An Introduction to Relativistic Quantum Field Theory*, Silvan S. Schweber, Dover Publications (June 17, 2005).
7. *Field Theory: A Modern Primer*; 2nd edition, Pierre Ramond, Addison-Wesley (1990).
8. *Gravity: An Introduction to Einstein's General Relativity*, J. B. Hartle, Addison-Wesley (January 5, 2003).
9. *Supersymmtric Gauge Field Theory and Strong Theory*, D. Bailin and Alexander Love, Taylor and Francis; 1st edition (1994).
10. *An Introduction To Quantum Field Theory*, Michael E. Peskin, Dan V. Schroeder, Westview Press; 1st edition (October 2, 1995).
11. *Beyond Standard Model*, M. E. Peskin, Proceeding of 1996 European School of High Energy Physics CERN-97-03, Eds. N. Ellis and M. Neubert.
12. *Vectors, Tensors and the Basic Equations of Fluid Mechanics*, Rutherford Aris and Mathematics, Dover Publications (January 1, 1990).
13. *Spacetime and Geometry: An Introduction to General Relativity*, Sean Carroll, Addison-Wesley (September 28, 2003).
14. *A Modern Introduction To Particle Physics*, Fayyazuddin and Riazuddin, World Scientific Publishing Company; 3rd edition (September 16, 2011).
15. *Statistical Physics*, L. Landau and E. M. Lifshitz, Butterworth-Heinemann; 3rd edition (January 1, 1980).

Printed in the United States
By Bookmasters